EYE *of the* STORM
Inside the World's Deadliest
Hurricanes,
Tornadoes,
and Blizzards

EYE *of the* STORM

Inside the World's Deadliest
Hurricanes,
Tornadoes,
and Blizzards

JEFFREY ROSENFELD

PLENUM TRADE • NEW YORK AND LONDON

ISBN 0-306-46014-9

© 1999 Jeffrey Rosenfeld
Plenum Trade is a Division of Plenum Publishing Corporation
233 Spring Street, New York, N.Y. 10013

10 9 8 7 6 5 4 3 2 1

A C.I.P. record for this book is available from the Library of Congress

Contents

Acknowledgments

This book began to take shape a dozen years ago in a (relatively) palatial maze of editors' desks, half-jokingly and half-wishfully named "The Ballroom," at Heldref Publications in Washington, D.C. When the day editor-superior Alice Gross introduced me to the man who would be my supervisor on a new assignment, I had no idea that I was in fact being introduced to a fascinating new culture. For five years Patrick Hughes and I would edit *Weatherwise* together while sequestered in a small room barely big enough for him to stretch his long legs across his desk. In that time and beyond, after Pat generously handed the reins of the magazine over to me, he practically channeled an appreciation for weather into my soul and poured into my head as much as could fit of his years of experience as a forecaster and weather savant. Pat made *Weatherwise* the educational experience of a lifetime.

The world he introduced me to—of clouds and winds and terrible tempests—is filled with legions of people who share a special affinity for, fear of, or insight into weather. Nearly all of them, at one time or another, came in contact with *Weatherwise*. They made a tremendous impression on me with their excitement and dedication. I could not help but mark the remarkable collegiality, good humor, and, yes, humility, that weather fanatics share in the face of seemingly overwhelming atmospheric chaos. Undoubtedly, these qualities make meteorology an unusually welcoming science. Not long ago, as I reread James Glaisher's excited tale of soaring to the unknown heights of the clouds, I realized that writing *Eye of the Storm* was my attempt to explain to the world how I came to know meteorology as a world of enthusiasts. I cannot imagine anyone who sat in that office with Pat and me remaining unmoved by the "sea of air" around us.

So I thank everyone who has written to or for *Weatherwise*, or has called to share, help, or even complain over the years. Their ideas and sentiments have driven me to my own unabashed excitement over weather and its science. In particular, Murray Mitchell and Bob Ryan, on their eagerly anticipated visits to our office, often helped clarify issues of significance. Dick DeAngelis gave a wealth of insights into weather and its history, as did the unforgettable writings and presence of the magazine's founder, Dave Ludlum, who I found, to my delight, also shared a love of history and matters orange-and-black. Mary Reed, with unfailing resourcefulness, proved to me the human richness of this subject. Al Blackadar, Craig Bohren, and Tom Schlatter, through their deeply and carefully considered ways and words, ignited my interest in the fundamental issues of science demonstrated by weather. The old issues of the magazine became well worn with use as I turned back again and again to their work as well as to contributions by dozens of other talented writers. As an editor, one could never hope to completely satisfy this dedicated, world-class group of experts. As a writer, I hope to honor their influence as best I can.

Many writers shared generously of their time and enthusiasm for weather—Nolan Doesken illuminated the special pleasures of snow, Dave Hickcox the delight of temperatures and small towns, Paul Kocin the unique variety of East Coast storms, Tim Marshall the puzzles of wind damage, and Lee Grenci and Doug LeComte pretty much anything that had to do with weather. Jack Williams also frequently shared generously his unparalleled perspective on the field. Master teachers like Ira Geer, Steve Richards, Kathy Murphy, and others, many of whom I met through the American Meteorological Society's vital Project Atmosphere, helped convince me over the years that the ideas in this book are worth the time and effort to share with people of all ages.

In the last year, as I focused on this manuscript, I had helpful conversations with such generous spirits as Joe Friday, Chester and Harriet Newton, Joanne and Robert Simpson, Hugh Willoughby, and Josh Wurman. Further interviews with Howard Bluestein, Roger Wakimoto, Morris Weisman, and Stephen Zebiak also spurred thoughts toward dozens of key questions in meteorology. Steve Horstmeyer provided a thorough and helpful review of the manuscript of the book itself, and Joe Golden also made suggestions that yielded crucial improvements. Any lingering errors certainly are my own doing at this point.

I also thank (and exempt) the two people who have been among the most helpful and supportive of my weather teachers, Stan Gedzelman and Bob Henson. Both took much of their own time to explain storms—Stan with his fabulous textbook, detailed articles, careful explanations, and amazing ability to ferret scientific merit out of photo contest entries; and Bob with his creative articles, delightful conversation, and memorable storm-chase tutoring as well. They certainly did their best to help bring me into the fold of weather-savvy writers.

As much as this book is about weather and my own love of the puzzles weather presents, it also is a labor of research and writing. This brings me to thank those who made it possible for me to actually write *Eye of the Storm*. The real beginning, and the continuing encouragement, comes from my parents, Robert and Linda Rosenfeld, who always helped me to follow my varied muses. I certainly cannot thank them enough for their support and advice. Their boundless help alone, however, could not have enabled me to write a book without the advice of many others: Through clear, dynamic, and dedicated teaching, Larry Mack began my training in his newspaper classes at Jackson (Michigan) High School. Alice Gross then brought me into professional publishing, rescuing the journalist in me after college; Sheila Donoghue fought battles for me and has been a valued source of encouragement; Barbara Kahn, Nanette Wiese, and Barbara Richman all provided professional wisdom, sage perspective, good humor, and warm companionship as I learned the ropes of the magazine business. Doug Addison and Doyle Rice, my editorial partners at *Weatherwise*, both gave me great advice, great ideas, and ultimately some of the best times I can imagine having while earning a paycheck. Merrill Joan Gerber graciously led me to Andree Abecassis, my agent, whose much-appreciated conviction in me and in the subject matter, in turn, led me to the wise and supportive Linda Regan at Plenum, where Vanessa Tibbitts, Nicole Turgeon, and Cathy Jewell all helped make these pages possible with hard work and helpful advice. Here at home, Susanna Spiro gave me comfort, understanding, and strength as together we endured a seemingly unrelenting succession of deadlines imposed by this year of writing and concentration.

But of the many who made an author out of me, Pat Hughes stands out again, for as he awakened me to the weather, so he also introduced me to the art of writing a good story. Editing my own first efforts and the work of others, he repeatedly polished rambling, raw text into anecdotal gold, always sharing the process with his usual generosity and irrepressible enthusiasm. In hopes that this book can honor in some way his inspiring example and valued friendship, I dedicate *Eye of the Storm* to Patrick Hughes.

CHAPTER ONE

A Stormy Relationship

It almost doesn't seem possible anymore. Not here, not in this century. It's only a storm story, but if you look carefully at what happened, it could have been an unthinkable political upheaval—a coup d'état, maybe a revolution. Here is how it plays out: On January 6, 1996, Congress and the President are exhausted from locking horns over a budget stalemate. Forces beyond our control seize the opportunity and bring Washington to its knees. Within two days, all but essential government employees are locked out of their offices on the very day their weeks-long furlough appears over.

In a few short days the State Department backlog of work mounts to over 100,000 unanswered visa applications. Not that anyone is going anywhere soon, anyway. Local governments in the Northeast are compelled to ban private cars from the roads. In many states, everything from side streets to superhighways is reserved for official use only. Drivers have abandoned their cars along the road where necessary. The airlines, tipped off about the impending crisis, have evacuated their planes from the tarmacs of coastal airports, stranding millions. Cancellations spread nationwide as hub after hub is shut down. By the end of the weekend, stranded passengers jam hotel suites and waiting lounges alike at airports as far away as Chicago, Denver, and Los Angeles.

In New York City—the city that never sleeps—the streets and stores are uncharacteristically quiet. Without traffic to bother them, amazed pedestrians find themselves walking down the middle of Manhattan's broad avenues. In Philadelphia, the *Inquirer* doesn't appear at people's doorsteps—a day without the paper for the first time in 166 years. People from Atlanta to Boston and beyond are reluctant to leave their homes. Some are running low on food or have lost electrical power. Parents keep their children home, and the elderly stay indoors. Listening to radios and televisions, people recognize the serious tones of emergency broadcasts. They hang on to every word as announcers drone on with lists of schools, governments, club meetings canceled—gatherings of every kind now

1

impossible. Reporters describe over and over the events leading up to the disaster. Analysts point to satellite photos on screen and show diagrams and 3D simulations of the crisis in action.

This was the scene—at least the grim face of it—during the Blizzard of '96, a storm that affected 100 million people, stopping them in their tracks and costing them perhaps $3 billion in damages and emergency costs alone. Nearly 200 people succumbed to the storm—in the cold and wind, in the ensuing floods, or even while attempting the arduous task of clearing four tons of snow from a driveway. After nearly a day and half of steady precipitation, 100 million tons of snow sits on the streets of New York City. The Big Apple is glazed with a crystal coating more than 20 inches thick. As far south as the uplands of Georgia, people are staring up at eight-foot drifts.

A large system of weather like the Blizzard of '96 brings more than snow: as the storm's core marched through the Northeast, winds in the most intense zones of activity howled to 50 m.p.h. Ocean waves driven by the long fetch, or distance, of wind over water, pummel Long Island—three houses and a restaurant fall into the ocean. The clearing sky and Arctic air mean bitterly cold nights following the storm. Temperatures will drop into the 20s in a few days as far south as Florida.

But blizzards are snowstorms, and it is the snow that ultimately shuts down the Eastern Corridor. In this case the snow is plentiful—both deep and widespread. More than 40 inches falls in higher elevations in West Virginia and nearly a foot in the mountains of South Carolina, while a record 14 inches buries Cincinnati. Two feet of snow falls in Rhode Island and Connecticut as well as in the Washington, D.C., suburbs. The 20.6 inches is the most in New York City since the famous storm of 1947. In Philadelphia, a record 30.7 inches falls. The city's plowing crews have nowhere to put it, and city officials plead with environmental regulators for permission to dump it on the frozen surfaces of the Schuylkill and Delaware Rivers. The snow itself not only paralyzes, it crushes: The roof of a Long Island supermarket collapses onto customers, injuring 10. A barn in Woodsboro, Maryland, also succumbs, killing more than 150 cows.

After the heavy precipitation buries its victims during the passage of warmer air, cold airstreams follow with strong northerly gusts that whip snow into whiteout conditions, endangering the hardy souls manning the great plows. The battle for mobility is lost as wind whips the drifts back across runways and roads as soon as the pavement is cleared. Because snow and ice-laden trees have already toppled onto power lines, millions of people are now huddled in the dark in poorly insulated homes as temperatures plummet.

To add insult to injury, just when roads are clear enough for government work to resume on Thursday, two more snow storms follow in rapid succession, adding nearly a foot more snow on Washington's hopelessly clogged streets. Two feet of snow falls in Pennsylvania. Inevitably, the barrage of low-pressure areas along shifting jet stream channels ends; a healthy patch of warm air liberates the Northeast between cold snaps. Burlington, Vermont, basks in 65-degree warmth at mid-month. With the sudden warming comes the potential for unseasonable thunderstorms: In Philadelphia 67-m.p.h. gusts precede pea-size hail. On January 19, the sudden changes of air also bring warm rains—one to three inches of it in New York and Pennsylvania. Waterwise, it's the equivalent of another foot or so of snow—but rainwater runs fast. The snow melts and ice dams crack, sending even more water down mountain streams into city river fronts. This time Wilkes-Barre, Harrisburg, and other Pennsylvania cities pay the price. At least 33 people drown and 51,000 homes are damaged; the state bears nearly $1 billion in costs from the storm and ensuing floods. Further along the water's course, the swollen Potomac threatens landmarks in Washington, D.C., tearing up urban parklands on the way to the sea.

The media trumpet the storm as the Blizzard of the Century. Weather experts have a bone to pick immediately: "blizzard" is a special word for special storms, and it is unclear if this storm fits the bill exactly. Very few, if any, weather observers among the thousands of volunteers who work with the National Weather Service report the requisite conditions: 35-m.p.h. or stronger winds and a quarter mile or less visibility due to snowfall over at least three hours. So maybe it was a blizzard and maybe not, but no expert will argue that the storm wasn't remarkable. More troubling is that such storms seem to happen with too great a frequency for unique historical claim. Just three years before, an even larger, more deadly, more powerful storm brought much of the East Coast to its knees so ruthlessly that meteorologists needed a new term, "superstorm," to describe it. The Superstorm of March 1993 dropped not only an incredible extent of deep snow over the most populous region of the country, but it also mustered a flank of thunderstorms that spawned tornadoes and 100-m.p.h. gusts as it swept over the Gulf of Mexico, past Cuba and Florida, and up the Atlantic coast. High seas off Canada and Mexico ripped apart merchant ships. Hurricane-like surges of water damaged coastal structures in Florida and the Carolinas.

It is tempting to call it bad luck. Why else would there have been yet another superstorm just one month before the Blizzard of '96 froze the East in its tracks? This storm smashed against the Pacific Northwest with winds gusting to 119 m.p.h. on the Oregon Coast; 28-foot swells dashed against the shore from San Francisco to the mouth of the Columbia River. Seattle and numerous cities in Oregon had never experienced a lower pressure than this hurricane-strong storm. Nothing like it had been seen since at least 1962, when the infamous Columbus Day "Big Blow" that flattened forests throughout Washington state.

But bad luck is not our problem at all. Storms always have, and always will, rule the roost on this planet. We are at their mercy. They are inconvenient, rude, dangerous, even deadly interruptions to life. An endless stream of intrusions—think of it: about a thousand tornadoes strike the United States in a typical year. One could argue that the more intricate,

A tornado rips through the landscape near Newcastle, Nebraska. Steve Liebetrau/*Weatherwise* photo contest.

fast-paced, and sheltered our lives become, the less we expect the intrusion of weather. We forget how unforgiving storms are.

Kublai Khan found out the hard way in 1274 when he decided to force the leaders of Japan to add their tribute to the coffers of his empire. He sent a massive force—900 ships bearing 40,000 of the finest fighters of the era—to the island nation. The Mongol cavalry, armed with poison-tipped arrows and arrayed in advanced tactical formations, began an island-hopping campaign of terror. At Kyushu, the invaders seemed destined to defeat the Japanese who rushed to defend their land. No records of the event survive, but legend has it a storm destroyed more than a quarter of the Mongol vessels in the harbor, forcing the invaders to evacuate and retreat to the mainland.

Kublai Khan had good reason to believe that the failure was a fluke. A deity in his own right, luck had never betrayed him so boldly before. So he gave the weather, and his warriors, another chance. The Khan ordered whole forests cleared in China to build new fleets. They set sail from Korea and China and arrived at Hakata Bay, Kyushu, in August—the height of the typhoon season. This time history is unambiguous: on August 15, two days of ferocious rains began. Furious winds and towering waves dashed the Mongol ships to pieces; the storm-stirred surge of water into the harbor made escape impossible. More than 100,000 Mongol warriors and their sailors were killed in the storm or abandoned to certain death at the hands of the Japanese. It was said that the wreckage of ships was so dense that a person could walk across Hakata Bay after the storm. The typhoon gave birth to a legend of invincibility in Japan: a divine wind, "kamikaze," was said to protect its shores from invasion.

In 1588, the Spanish would relive Kublai Khan's frustrations. That was the year King Philip sent his Armada to conquer another island nation—England. The Spaniards intended to dispose of the British fleet to pave the way for their landing party of 30,000. The English Navy was rather scrappy by its later, polished standards, but their ships were fast and their sailors adept in home waters. The defenders sank one Armada ship and captured several more. By superior maneuvering in shifting winds, the English were able to get a few Spaniards to run aground in Flanders. In other words, Spain's invasion plans were profoundly inconvenienced by battle. But they were not defeated—merely put off for the time being. It had taken a month just to sail within range of the English fleet, and the Spaniards were low on food and morale. They were determined to return to Spain. That's when storms intervened. Fifty of the Armada's ships—

nearly half the remaining fighting force—sank off Scotland and Ireland as they met some of the worst summer and fall storms ever recorded in that part of the world. Most of the 6,000 Spaniards who died in the campaign were lost to those storms. The weather had determined the course of Spain's empire—not to rule England or correct religious practices there, but to frantically mine gold to pay off the tremendous debts that the ill-fated Armada incurred. Like the Armada, the ships bearing Spanish gold would litter the bottom of the sea. For centuries hurricanes were Spain's most formidable enemy in the New World.

Our relationship with storms has changed considerably since King Philip's time, even if our luck hasn't. We still get storms, and it's a given that we get very big ones, very powerful ones, very frequently. Now, however, most of us live and work comfortably in well-built, heated buildings. But this comfort isn't the whole story of our changing relationship with the weather. Nor is the story found in the seawalls we build to protect our coastal cities, or the cars we drive through rain and hail, or the airplanes we fly above the clouds. We have all of these and still freeze in

A menacing wall formation lowers from the base of a towering thunderstorm, signaling a churning rotation at the heart of the clouds. David Thede/*Weatherwise* photo contest.

cold weather, get stuck in mud and snow, and watch our beaches erode in the waves and planes crash in thunderstorms.

Our relationship with storms has been shaped by an amazing story of scientific progress in our time: the stratospheric rise of meteorology. You can mark the change in meteorology in many ways: a century ago there were no universities in this country offering a degree in meteorology. Today there are nearly 100 such programs. There were no university-educated meteorologists. Now, to become one of the 10,000 full members of the American Meteorological Society, you need to have completed the equivalent of a bachelor's in meteorology. Weather forecasting was once a laughingstock. Its practitioners considered their work an art, not a science, and their insights died with them as secrets in the trade. Today the field has nearly a dozen top-flight journals devoted to various aspects of the science, and forecasts are scrutinized and improved by rational, mathematical means.

But these are the insider's gauges of success. Meteorology is a science that openly shapes the way we live. Through forecasting, it is a most public of sciences, displaying its successes, and failures, in our newspapers every day. No scientists except meteorologists ply their trade in front of millions of viewers on newscasts every day.

The real mark of meteorology's rise to the front rank of science is apparent in the way we live with storms today. And it is apparent that our relationship with storms has changed drastically as scientists have remade our understanding of weather in a very short time. The Blizzard of '96 shows that progress of understanding, and the attendant shift in perspective. At the same time it shows we remain profoundly vulnerable to storms despite the knowledge we have acquired.

Yet consider the incredible progress meteorologists have made. Just 13 years before, in 1983, Washington, D.C., was inundated with more than two feet of snow in a similar, widespread blizzard. Outlying areas had more than three and four feet of snow. It was as "can't-miss" a forecast as any huge East Coast snowstorm, and yet citizens had little warning that time. The supercomputer forecasting models at the National Meteorological Center had a great deal of trouble predicting fast-developing storms back then; worse yet, coastal snowstorms are a highly sensitive problem to simulate. In 1996, however, meteorologists at the center in Camp Springs, Maryland, were running a much better computer model, on a much faster computer. The results, as during the 1993 Superstorm, were stunning. On Thursday, January 4, meteorologists became convinced snow would start falling on Saturday night and end some time Monday—at least twice as long as a typical snowstorm. On Saturday and Sunday, they slept over at the office fully aware of the kind of week ahead. They got the word out that this would be a huge snowstorm, and the warning paid off. Before a single snowflake had fallen, Governor George Allen had called out the National Guard in neighboring Virginia. Residents scoured grocery stores for supplies, and at least one parent was pleased she could plan ahead by buying a toboggan for her children, who'd be homebound the following week.

Here's another way to look at the fast pace of weather science in this century. In 1938 a hurricane off the Florida shore looked like it might hit near Jacksonville. As citizens there boarded up windows in preparation for the worst, Weather Bureau forecasters received radioed ship reports claiming that the hurricane was turning north and northeast. This is a typical recurvature of a tropical storm's path over the Atlantic—from westward, to northward, to northeastward around the backside of a big area of high pressure that lingers around Bermuda. All signs were that the coast was clear, and Easterners breathed a collective sigh of relief. Unfortu-

From airplanes we gain a new perspective on thunderstorms. NOAA photo.

An October 1997 blizzard brought Colorado to a standstill with several feet of snow. Mark Reis/*The Gazette* (Colorado Springs, CO).

nately, ship reports were the only means for the Weather Bureau to track these storms, and when a hurricane is imminent, smart sailors steer clear. The next day, people on the Long Island shore scanned the ocean and saw what looked like a thick layer of gray fog rolling in toward them. A few minutes later they realized to their horror that the fog was in fact a 20-foot surge of hurricane-driven water. Overnight, unbeknownst to anyone in the Weather Bureau, the storm had suddenly accelerated to 60 m.p.h. and barreled its way across the Atlantic like an express train. Dubbed the Long Island Express, the storm killed hundreds of unprepared people on the beaches of New York and Rhode Island, flooded downtown Providence, and flattened millions of trees in northern New England.

The Blizzard of '96, by contrast, was a closely monitored storm. Satellites operated by the National Oceanic and Atmospheric Administration, the parent office of the Weather Service, kept a constant watch on the development of the clouds as well as an eye on the invisible rivers of water vapor streaming into the storm, making it a veritable snow machine. In 1938, had forecasters viewed a map of atmospheric conditions at about 18,000 feet, they would have noted the valley of low pressure inviting the hurricane straight in to the Northeast. In contrast, regular balloon soundings of the upper atmosphere in 1996 were tracked by radar and reported temperature, moisture, pressure, and winds. Whereas theories about upper-level flow were brand-new and largely untested in 1938, in 1996 the balloon data clearly showed the steering path marked for the blizzard as well as the regions of most intense winds, and thus energy, that were moving over the storm just as it hit the mid-Atlantic states.

If you make a quick tally of the technology behind the boom in storm science—computers, radar, satellites (particularly the rockets that launched them)—you can detect the source of recent meteorological development: World War II. Meteorologists were among the first scientists to capitalize on the tools of victory. As calculating machines advanced from punch cards to parallel processing supercomputers, meteorology grew from guesswork with hand-drawn maps into a science of cybersimulation. As airplanes went from cloth-wrapped kites to sturdy jets, meteorologists ventured into the unknown reaches of severe thunderstorms and hurricanes. As rockets leapt from fireworks to space-bound behemoths, meteorologists added a whole new perspective to their view of storm structures. Moreover, the widespread use of radio gave meteorologists a chance to gather critical data about the upper atmosphere; the radio waves in radar gave meteorologists a chance to witness tornadoes, snow, and hail grow within the clouds.

In World War II, universities trained an unprecedented number of men and women in weather sciences. These thousands of technically trained storm experts proved critical to the success of naval operations, bombing raids, and, in particular, to amphibious assaults. D-day itself hinged on its weather forecast. Eisenhower called back his invading force from the sea because of bad weather and consulted with his meteorologists in England. When did they expect a break in the rough weather? The general needed very specific conditions: low coastal cloud cover for gliders and paratroopers to drop unnoticed into France at night, good visibility later for bombers to spot targets, high tides early for submarines to prowl close to the shoreline, and calm seas at low tide for weapons-laden youngsters. Eisenhower had competing teams of British and American forecasters in two different offices, using different methods. Finally they agreed that an opportunity for appropriate weather would arrive on June 6, and Eisenhower grabbed it. Significantly, German meteorologists had mistakenly written off invasion prospects that weekend because they did not expect the bad weather to let up. It was a meteorological blunder, then—perfectly understandable with the few observations available to the Germans—that enabled the Allies to catch their enemy off-guard.

But the thousands of meteorologists trained in World War II were also acutely aware of the limits of their skills and tools, and at no time more so than when the Navy's Third Fleet cruised the Pacific in December 1944. The fleet, commanded by Admiral William F. "Bull" Halsey, was the mightiest battle group ever assembled at that time. The 12 aircraft carriers, nine battleships and more than 70 cruisers and destroyers and their support craft maintained a formation covering 3,000 square miles of the Pacific. On December 18, as Halsey's ships steamed toward the Philippines to support General MacArthur's landings, meteorologists aboard became anxious. Two days before, observers on both Guam and Ulithi had reported force 4 (13–18 m.p.h.) winds—in opposite directions. It was a subtle clue, but American forecasters ("aerologists" in the Navy lingo) understood: in between was a swirl of wind around low pressure—probably a growing typhoon. On Halsey's flagship, the *New Jersey*, pressure fell steadily on the morning of the 18th, indicating the approach of a storm. But no one in the fleet knew precisely from where. Halsey consulted his chief meteorologist, George Kosco. The MIT-trained expert had little information to go on but the vagaries of wind and pressure, much like the buccaneers of centuries past who had first deduced the circular nature of these tropical whirlwinds. Rough seas had made aircraft reconnaissance impossible— refueling operations had already nearly led to ship collisions. Far from Kosco, aboard the carrier *Wasp*, radar operators were puzzled by a mass of white reflections on their screen—they didn't yet know that the new instrument was as useful for monitoring the weather as it was for tracking enemy planes.

But Kosco did have at least one leg up on the mariners of previous centuries. He knew from climatological averages that, historically, three-quarters of the typhoons in the region passed northeast of the *New Jersey*'s position anyway. He guessed the storm was 200 miles to the northeast of the *New Jersey* and moving away slowly. Thanks to the discovery of weather fronts during the previous war, he was also aware that cool Asiatic air—a cold front—was driving under the Pacific atmosphere and causing choppy seas and worsening skies. Halsey, ever the gutsy admiral, had to make a decision. Perhaps he was a bit too daring, having just a month earlier maneuvered behind the mask of another Pacific typhoon to surprise the Japanese. Halsey figured his sprawling fleet had enough sea

Tornadoes are nature's most horrific winds: a man contemplates the ruins of his home in Moody, Texas, after a May 27, 1997, outbreak that killed 29 people in nearby Jarrell and Cedar Park. Duane A. Laverty photo.

room to ride out the fringes of a storm and a cold front before beginning support of MacArthur's troops. After a 5 a.m. briefing on the 18th, he chose to turn south.

It was an unfortunate choice. Guts, and certainly climatology, are a moot point when the reckless, capricious fury of a storm is at hand. The typhoon, later nicknamed Cobra, turned out to be only 90 miles to the east-southeast. Cobra charged northwest, right into the fleet's new course. Kosco and his staff watched in horror as the mercury dropped two-tenths of an inch around 10 a.m. The full fury of the storm's central cloud bands became apparent over the next four hours. The sea spray was so thick over the *New Jersey*, Halsey couldn't see the ship's bow from his bridge. The mighty battleship shuddered from watery blows that felt more forceful than a Kamikaze crash or an exploding shell from the guns of an enemy cruiser.

Smaller ships fared much worse in the storm. In the 110-knot winds, one destroyer slid across the wave crests as if on skis, then disappeared momentarily into a veritable canyon of water. Churned in the belly of the storm, the destroyers rolled side to side until their stacks were nearly horizontal. One destroyer escort crew figured they wouldn't recover from the rolls because their communications tower kept dragging in the water, so they cut it off with torches. Other crews didn't have a chance. Low on fuel, the already top-heavy *Farragut*-class destroyers tipped precariously in between the walls of water 60 feet high. Water rushed below decks, frying electrical circuits. Three of these destroyers disappeared in the raging water, and with them 790 men met their deaths in the deep ocean.

Experiences like these convinced many of the finest wartime scientists to stick with meteorology. Many of the best and brightest found themselves attracted to storms. They had seen the sleek new planes and rockets and radars during the war, and figured all these advances could be put to good use at home. The ideas about storms explored since World War II are in part a result of the new vistas opened up by technology. But these ideas about storms also have many roots in the first half of the century, even in the 19th century.

The device with the most profound impact on the methods and capabilities of meteorology was the computer, and not just because computers now make storm forecasts or process Doppler data from radars and generate images from satellites. Almost by itself, the computer has expanded meteorology from its humble beginnings as an observations-only science. With a few exceptions (such as precipitation experiments), weather has defied the controlled conditions of a laboratory that confer on physics, chemistry, and biology a special status. The advent of cyberspace has offered nearly unlimited possibilities for testing ideas about weather. Even the digital overhaul of storm science in the last 40 years, however, was anticipated in part by the introduction of thermodynamics in meteorology in the 19th century.

The first half of the 20th century in particular saw profound discoveries in meteorology: a new model of how a typical midlatitude storm works, a fresh understanding of the role of storms in the general circulation of the atmosphere, a convincing theory of how rain forms, new theories about tornadoes, and more. The growth in aviation spurred on much of this work in our century. To a nearly equal degree, reliable air travel, a marquee achievement of our age, owes much to our improved understanding of storms.

Two brothers indeed overnight made meteorology a contender among heavyweight sciences. Not the Wright brothers in 1903, necessarily—but instead the Montgolfier brothers in 1783. Scientists were among the first to hop in for a ride in the Montgolfier's new hot-air balloon, just as scientists were among the first in space in our day. With the balloon they charged through the clouds and wondered at the appearance of grand storm systems. In the air they carried thermometers, barometers, and other instruments to confirm the laws of chemistry and physics that would reshape meteorology. They met storms on their own terms for the first time. Writing in the *New York Tribune* in 1857, aeronaut John Wise tells of a rough ride he took in a thunderstorm over Carlisle, Pennsylvania, in the rickety basket of a hot-air balloon. He heard the "howling noise" of thunder, smelled a sulfurous odor once commonly attributed to lightning, and saw with his own eyes the sheets of water suspended in air. "The color of the cloud internally was of a milky hue," he reported, "somewhat like a dense body of steam in the open air, and the cold was so sharp that my beard became bushy with hoar-frost.... Little pellets of snow ... were pattering profusely around me in promiscuous and confused disorder."

Before getting drenched in the rainshaft below the cloud, Wise's balloon shot up like an elevator in a skyscraper. It was a visceral confirmation of the hot updrafts basic to severe weather that, a few decades earlier, were beyond the imagination of scientists obsessed with horizontal winds:

> [S]light blasts of wind seemed occasionally to penetrate this cloud laterally, notwithstanding there was an upmoving column of wind all the while. This upmoving stream would carry the balloon up to a point in the upper cloud, where its force was expended by the outspreading of its vapor, whence the balloon would be thrown outward, fall down some distance, then be drawn into the vortex again to be carried upward to perform the same revolution, until I had gone through the cold furnace seven or eight times....

Behind the success of this dynamic pairing—the exploration of the sky and our newfound acquaintance with storms—is an even broader, quieter sea shift of meteorology. These are the grand trends of human perspectives over the last three or four centuries that broke the back of thousands of years of superstition and emboldened the early aeronaut/ scientists to explore the weather. In a few pages you will read about one man in colonial America who broke the spell of ignorance that hung over storms until our times. But here it is worth seeing some of the trends supporting his earth- (and weather-) shattering insight.

The change in the scientific and popular approach to storms first stirred in the meteorological numbers racket that began with Galileo, who

Storm surge from Hurricane Carol in 1954 inundates the Rhode Island Yacht Club. Carol was the costliest of five landfalling hurricanes in the Northeast in two years, a spate of disaster that led to a rapid expansion of the Weather Bureau's radar network. NOAA photo.

invented the thermometer in 1607. Later, in his dying days, he suggested a tubular contraption that might prove the ever-present weight of air in the atmosphere. His assistant, Evangelista Torricelli, took him up on the idea and in 1643 invented the barometer. It didn't take long for astute observers to link the fall of local air pressure to stormy weather. Not by accident did Torricelli write, in a letter to a friend in 1644, "We live submerged at the bottom of a sea of elemental air."

It is true, of course, that little scientific use came of the new instruments before meteorologists organized networks of trained observers. But the future direction of meteorology was set. Meteorologists for centuries have been assiduously collecting data, hoping for a tool to process it all. It is as if the words "computer" and "gigabytes" (and now "teraflops") were coined just for them. You can see the weather data obsession in the diaries of astronomer Johannes Kepler, a contemporary of Torricelli's, who kept daily records of passing storms, and later with such devoted weather diarists as two gentlemen farmers of Virginia, Thomas Jefferson and George Washington. Jefferson even proposed data exchanges with his correspondents and insisted that Lewis and Clark make careful notes about the weather of the West on their journey.

At the beginning of the 19th century, meteorology was caught in the classification rage that reignited the natural sciences. One of the most unlikely innovators of this era was a young man named Jean Baptist de Monet, later known as Chevalier de Lamarck. Born in 1744 to a noble family that committed him to church at an early age, young Jean hungered for adventure and rebelled against his family's plans. In 1761 he joined the army. Lamarck won honors on the battlefield, but his auspicious military career was short-lived. Injured while roughhousing with his companions, he was sent to Paris for recuperation. Confined to a room with a small window overlooking the roofs of the city, Lamarck had little entertainment but the passing cloudscapes. He decided to become a scientist, devoting most of his years to a classification of species that would rule the natural sciences for many years.

He also applied his penchant for classification to the clouds he had watched while recuperating in Paris, and in 1802 put forth a less remarkable, but still revolutionary, categorization of clouds. Clouds, he wrote, were not merely a panoply of endless variation: "[I]n addition to the special and accidental form of every cloud, one clearly notices that clouds display certain common forms, which are in no way accidental, but which can be usefully recognized and distinguished...." Lamarck determined there were five basic forms: scarf clouds, checkered clouds, fleecy clouds, swept-up clouds, and clouds in group form. The last category meant storms. He wrote:

> The grouping,... huddling together and piling up of clouds, which makes them—when seen from the side of the sun—look like heaped-up mountains and which gives them a considerable height, is in fact important for the progress of meteorology. I think one will never get a clear and sound idea of the causes of thunderstorms if one does not pay attention to the theory of grouping of clouds and to the consequences of this state of grouping....

This sentiment still drives much research into clouds in all kinds of storms, not just thunderstorms.

It was a sign of the careful attention to description and categorization in those times that only one year later a remarkable English chemist, Luke Howard, devised the cloud chart scheme that persists to this day in slightly modified form. Howard, born in 1772, shared Lamarck's interest in botany, and in fact his first scientific paper was on pollen. At the same time he had a boyhood interest in weather, which flourished. He bragged about his early obsession with the barometer. Eventually he saw fit to publish his daily meteorological journals in a popular magazine of the day.

Howard's profession as a chemical supplier provided him with a scientific background. In his spare time he set up a home laboratory for electrostatics to duplicate the experiments of the legendary Benjamin Franklin. He befriended John Dalton, who formulated the first scientific theory of atoms. In 1837 Howard published *Seven Lectures in Meteorology*, the first comprehensive scientific meteorology textbook.

During the winter of 1802–3 Howard read his "Essay on Clouds" to a small, scientifically minded society of his friends. He proposed three basic categories, or "modifications," named in Latin (an exercise of traditional grammar school values that Howard actually despised!): First there were *cirrus*, made up of extended "parallel, flexuous, or diverging fibres." Next, *cumulus*, or "convex or conical heaps" with a "horizontal base." Finally, *stratus*, or sheet clouds. In between were intermediate and hybrid categories, including cirrocumulus ("small, well-defined roundish masses" in decks), cirrostratus ("horizontally or slightly inclined masses" that tapered on edge), cumulostratus (combining cirrostratus with lumps of cumulus), and "cumulo-cirro-stratus vel nimbus" (or rain clouds, largely in sheets). Howard didn't expect these categories to remain unchanged, and indeed he tinkered with them in his own lifetime. Today, however, our categories are still based on Howard's distinction between lumpy, stringy, and deck clouds. A major addition, the separation of these shapes by altitude (yielding categories such as strato-, alto-, and cirrocumulus) emerged many decades after Howard's death.

In the summer and fall of 1803, Howard's scheme was published in installments in England and shortly thereafter was translated into French and German. One German translation inspired Goethe to write a series of poems—"Cirrus," "Cumulus," "Nimbus," "Stratus," and later, "To the honoured memory of HOWARD." The German author began a correspondence with Howard, and more importantly, promoted Howard's scheme throughout Europe.

Howard expected that an understanding of cloud shapes would lead to scientific forecasting. And for many years, observations of the clouds, correlated with pressure maps and temperature changes, would inform many a cyclone theory. It took more than a century before meteorologists

Lightning still kills 100 Americans a year. Here a bolt forks into the ground in Connecticut. Arthur Stewart/*Weatherwise* photo contest.

reached Howard's goal with mathematical forecasts. In Howard's time, such a development was still impossible. Scientists had not yet articulated the physical principles shaping the atmosphere. Howard's cloud families, however, did finally bring structure to observations. Weather watchers could record the state of the sky and know that their observations would be intelligible to their colleagues and their descendants, not to mention their next-door neighbors. The categories worked their way into popular consciousness. Landscape artists like John Constable were inspired to paint the sky with new faithfulness, and today popular weather "field guides" remain largely vehicles for spectacular photographs illuminating the scheme initiated by Howard.

Weather categorizations are still with us, still evolving. We hear about Category 1 through 5 storms each hurricane season. We hear the Fujita scale of tornado strength bandied about (and misused) in blockbuster movies. These are fairly new, but less conceptually significant than even Lamarck or Howard's categorizations. More crucial categories are being debated in the meteorological research right now. The ground continues to shift over old questions. Scientists refine the distinction between a tropical

A funnel cloud reaches for destruction over Denver. Eddie Grubb/ *Weatherwise* photo contest.

storm and a midlatitude storm. They try to squeeze into their scheme the arctic storms that look like hurricanes on satellite imagery. They try to define what distinguishes a dust devil from a landspout from a waterspout from a tornado.

The technology of the modern age, the boldness of airborne science, the time-honored obsession with data collection, the precision of categories— all of these trends help us perceive the Blizzard of '96 as a living cloud family evolving because of powerful forces explained by elegant physics. And this is no arcane feat. Meteorology has its inaccessible, mathematical arena, but many of the new ideas about storms are aesthetic, visceral, and accessible. Dazzling satellite and radar imagery of the storm delighted Washingtonians engrossed in the continuous weather reports broadcast on their local news. The local NBC affiliate, with the help of NASA, had set up a popular web site about local weather. Technophiles visited it at a rate of more than 30,000 an hour during the height of the blizzard.

The storm, we could all see, had its roots in an outbreak of dense, cold air in New England. The blob of air sat undisturbed by fast moving upper air, while a storm formed off the Rocky mountains. At first the storm was a rather dry system; then the new low-pressure area rode the steering currents out over the Gulf of Mexico. Like a typical, incipient nor'easter, it picked up warm moisture there, forming a cloudy mass that began sailing northward over the southeastern states. At a critical juncture, over the Carolinas, the low-pressure area passed the torch to a new, more powerful area of converging air just offshore over the energy-laden waters of the Gulf Stream. Computer models had predicted that this translation would occur and that the new storm lacked a strong push from above and wouldn't move fast. The result was a record-breaking snowstorm. Satellite photos on our television screens showed the storm make its snow. West and north of the storm's center, long streets of clouds curved counterclockwise over the Atlantic, marking the maritime air streaming toward cold air hugging the New England terrain. The wet air rose up over this invisible mound, cooled by the laws of physics and the loss of pressure attendant with ascent. The rising air began to condense into a thick mass of cloud. Inside it, snowflakes grew and fell. When upper-air patterns changed, the storm finally moved north and ended, leaving behind a nation of frustrated shoppers, helpless travelers, and delighted schoolchildren.

Just like a championship football game, the blizzard was in part our entertainment: we knew about it before it would happen, we just didn't know how it would turn out. We watched it play by play, critiquing it with great skill. The satellite replays shown over and over on television helped us see when the cold air eventually pushed the warm air back out over the Atlantic. The science of storms has turned the weather into a vast playing field of physics and natural wonder. This is the scientific excitement that leads some meteorologists to hop in a car to try and drive to the exact spot where snow might begin to fall. They hunger to feel the action as well as to verify their instincts about the outcome. A few people are even turning storms into a business. While scientists and amateur storm chasers hunt tornadoes on the Great Plains to provide new data, new insights, and better warnings, you can actually now book yourself with some storm chasers for a few weeks' vacation to hunt for the big funnel.

That's the real revolution in our stormy relationship with weather. Thanks to an amazing century of scientific discovery, we're experiencing a veritable coup d'état of science over weather. In the new regime, scientists do not scheme to control the weather, as in the heady manipulations of the 1960s and 1970s, but instead provide us with a deep understanding of, and appreciation for, storms. It is our profound luck to live in a stormy era guided by great meteorological discoveries.

CHAPTER TWO

A Light in the Dark

"Good God! What horror and destruction!" wrote 15-year-old Alexander Hamilton to his father about a devastating hurricane in St. Croix in September 1772:

> It seemed as if a total dissolution of nature was taking place. The roaring of the sea and wind, fiery meteors flying about in the air, the prodigious glare of almost perpetual lightning, the crash of falling houses, and the ear-piercing shrieks of the distressed were sufficient to strike astonishment into Angels. A great part of the buildings throughout the island are levelled to the ground; almost all the rest very much shattered, several persons killed and numbers utterly ruined—whole families roaming about the streets, unknowing where to find a place of shelter—the sick exposed to the keenness of water and air, without a bed to lie upon, or a dry covering to their bodies, and our harbors entirely bare. In a word, misery, in its most hideous shapes, spread over the whole face of the country.

Hamilton's eloquent letter caught the attention of influential men in the American colonies. They brought the promising youngster to this continent for schooling, and he rose to become the new nation's first Secretary of the Treasury 17 years later. The sensitivity and objectivity of young Hamilton contrasted sharply with the sometimes judgmental tirades that storms provoked from men of previous generations. Wrote Increase Mather, the influential Puritan of late 17th-century Boston, "It is not heresie to believe that Satan has sometimes a great operation in causing thunderstorms." In an exhaustive study of New England disasters—of the "solemn works of Providence"—Mather detected an ominous consequence of deteriorating morality in the colonies: "[A]lthough terrible lightnings with thunders have ever been frequent in this land, yet none were hurt thereby (neither man nor beast) for many years after the English did first settle in these American desarts [sic], but that of later years fatal and fearful slaughters have in that way been made amongst us, is most certain...."

With storms, as much as with the chaotic, paint-splattered canvases of abstract art, what you see is what you think. Sometimes people seem to see nothing at all—it was extreme even in 1652, however, for the Reverend Randall Sillito of Cheshire, England, to continue delivering his sermon from his pulpit after a brilliant light temporarily blinded him and a blast like a "grenado" rang in his ears. Hearing no cry from his congregation, he had no idea that 11 men and women slumped over in their pews dead, incinerated by a flash of lightning that had hurtled down the bell tower. When Sillito finally noted the casualties among his parishoners, "some as though their feet and arms had been on fire," he seemed undisturbed that the atmospheric fury tested the passionate piety around him: "They that had friends carried them out in much silence, and we continued in preaching and in prayer about the usual time...."

A century later, it was still possible to see storms through the fog of religious fervor. It was also possible to see storms as a game for gods and politics, a long tradition upheld by the ancient Romans, who used certain patterns of lightning as a pretext for adjourning the Senate. But by Hamilton's time, a person of signal perceptiveness understood weather in a new way—a way still comfortable today. When Benjamin Franklin sailed

The long reach of the gods? Lightning can surprise its victims by striking a dozen miles from its origins in the clouds. James R. Doughty/*Weatherwise* photo contest.

to England in 1757, his ship nearly foundered on the rocky coast near Falmouth. This close brush with disaster couldn't shake his relentlessly practical nature. "Perhaps I should on this occasion vow to build a chapel to some saint," he wrote in a letter to his wife, "but … if I were to vow at all, it should be to build a lighthouse."

This attitude would have made Franklin a remarkable man in any era. In his own day, however, Franklin was a visionary and a cherished icon. Loyal readers consulted Franklin's *Poor Richard's Almanack* for homespun wisdom. Many more people benefitted from his practical inventions: because he felt burdened by too many spectacles, he devised bifocals; because he saw that streetlights quickly dimmed from soot, he improved the flow of smoke in lamps. He organized the first free library, the first public hospital, the first citizens' fire-fighting company, the first street-cleaning service, and much more.

The lightning rod was Franklin's greatest innovation, even more famous and universal than his ubiquitous stove. It saves thousands of lives and homes every year, which certainly ensures its place in history. But the lightning rod also represents a crucial development in the history of ideas. It literally transformed humanity's relationship with storms. The lightning rod was a huge intuitive leap in science and an ancient dream come true in meteorology. It was a light that brought meteorology out of the dark.

Admittedly, Franklin would not be a model scientist today—certainly not in meteorology. He had only two years of formal education and would not be very comfortable with today's computer models. Franklin didn't fully understand Newton's important treatise on gravity; the math was too advanced. (Franklin was in good company—John Locke, the great Newtonian philosopher, also admitted he couldn't read it.) But Franklin instead had a gift for unfettered curiosity, the mark of a great scientist. He tested various clothing in the sun because he wanted to establish a relationship between color and heat absorption. He fished out the flies floating in his wine to see if they were dead or merely drunk (most were the latter).

His delightful inquisitiveness made Franklin the foremost scientist of his day. Curious about the sluggishness of westbound mail on the Atlantic, he discovered and charted the Gulf Stream, the narrow band of warm water flowing north off the East Coast of the United States. Passionate about electricity, Franklin defined positive and negative charges and built batteries. Franklin's electrical principles made verifiable sense out of a

"Benjamin Franklin Drawing Electricity from the Sky," painted by Benjamin West around 1816. Philadelphia Museum of Art: Gift of Mr. and Mrs. Wharton Sinkler.

mysterious force known for thousands of years. For these discoveries, chemist Joseph Priestley unashamedly ranked Franklin with none other than the great Newton. Yet as important as Franklin's electrical insights were in physics, he was probably nowhere more significant than in meteorology. The modern study of storms began with Benjamin Franklin.

Franklin didn't actually devote himself to full-time science until 1746, after making his money as a printer, publisher, and postmaster. Already 40, Franklin had reached an age at which some scientists consider themselves over the hill. But Franklin had caught the electricity bug only three years before, on a visit to his native Boston. There he met the itinerant scientific lecturer Archibald Spencer, who tried to demonstrate for Franklin the amazing feats of the Leyden jar, a crude electrical capacitor poorly understood (before Franklin explained it) but widely used to make experiments. One of Spencer's most popular tricks was to suspend a boy in the air with silk straps hung from rafters and then rub glass tubes nearby. Static charge sent sparks flying between the spheres and the child. The experiments, if not Spencer's execution of them, impressed Franklin enough to sponsor the lecturer and his show in Philadelphia. "[The demonstrations] were imperfectly performed, as he was not very expert," Franklin wrote, "but being on a Subject quite new to me, they equally surpriz'd and pleas'd me."

Soon afterward Franklin began experimenting on his own with equipment sent to him by Peter Collinson, the London agent for his Philadelphia library. Later Franklin bought Spencer's own apparatus when the lecturer traded life on tour for the security of a Maryland pulpit. Franklin experimented with the static charges at home. His neighbors were so delighted that he presented them with glass spheres so they could make their own sparks.

The sizzling, leaping static sparks that mesmerized Franklin and his friends led to a magnificent discovery about the much-feared lightning bolt. The ancient Greeks had long ago recognized static electricity; they generated it when they rubbed amber with cloth to make it cling to other objects. (Appropriately, the word *elektron* is Greek for amber.) Yet, until Benjamin Franklin turned his prodigious talents to science, very few people considered the similarity between lightning and electricity. In 1708, William Wall noticed that rubbing amber created a light and crackling similar to lightning. He wrote about it in the *Philosophical Transactions of the Royal Society*, an influential scientific honor society begun by Newton and his circle. Nine years later, Newton himself contemplated the sparks generated by rubbing amber with silk, and wrote, "Ye flame putteth me in mind of sheet lightning." In 1746, the year before Franklin began his own experiments, J. H. Winkler published a treatise on electricity. "Are flashes and sparks of intensified electricity to be considered as sort of thunder and lightning?" he asked, answering with slightly twisted reasoning, "[T]hunder and lightning ... [are] ... the same as what is observed in the case of electric sparks. No sooner does lightning flash through air than the particles of an electric spark sweep into its body."

Yet all of this was idle speculation before Franklin. And for thousands of years idle speculation had been the bane of progress in the still very unscientific study of weather. Many decades into the Newtonian age, astronomy flourished under the aegis of the new physics; meteorology floundered. No one at that time would have been called a "meteorologist"; in the 18th century people dabbling in physical and geographical knowledge related to the atmosphere called themselves "natural philosophers."

But Benjamin Franklin, who was also nearly everything else, was a bona fide meteorologist. With Franklin, humanity quickly made up for more than two millennia of basic neglect of storm science. Our understanding of storms underwent a revolution more shattering than the one Franklin and his neighbors sprang on the world's mightiest empire. Because of Franklin, errant—even dangerous—speculation gave way to understanding. Franklin proved that storms could be studied with science. He opened the door to a new age of meteorology—to a quest based on confronting the violence of nature, on careful observation, and on theory based on physical laws—even if he left for others the job of showing what most of those laws might be.

This great awakening of storm science seems even more remarkable when you consider that the meteorology Franklin inherited had been dormant for about 2,000 years. The tenets of Greek natural philosophy had been improbably propagated across millennia of human history with only a few embellishments and many layers of superstition. Ideas about what the atmosphere was, how it worked, and what caused it to erupt into stormy fury—these theoretical underpinnings of meteorology still reflected the work of the man who invented the discipline and who, simultaneously, plunged it into a dark age of its own.

That man was Aristotle, born in 384 B.C. A student of Plato, Aristotle's fame as a natural philosopher grew quickly. Eventually he moved on to tutor young Alexander of Macedonia. Around 350–340 B.C., not long before he returned to Athens, Aristotle wrote *Meteorologica*, one of his influential treatises on the nature of the world. With this imaginative work, Aristotle gave birth to the study of the atmosphere. Unfortunately, he nearly killed it as well.

The reasons for the neglect of storm science are complex. Rather than investigate weather on their own, Seneca and Pliny, the major Roman commentators on meteorology, simply summarized *Meteorologica* and other Greek works in thick handbooks of knowledge. Others who took up Greek interpretations of weather included Virgil and Lucretius. The consensus of such revered authorities likely scared off many would-be meteorologists. Then, in the Dark Ages, "European science" practically became an oxymoron; the center of Western learning and inquiry shifted to the Middle East. Editions of Greek scholarship were translated into Arabic and Syriac languages. Some Arabs ventured beyond Aristotle's system to explain their own observations of weather but with little lasting effect on science. However, through the faithful summaries and translations by the Arab scholars, Aristotle's ideas survived to see the light of a new era. With the Renaissance in Europe came intense interest for all things Greek and Roman, including Aristotle. More than 100 editions of *Meteorologica* were published by the end of the 16th century. These books were second-hand reconstructions—the original having long ago disappeared. By this time, Aristotle's ideas were closely allied with Church doctrine. Christians were not alone in embracing Aristotle, of course. The medieval Jewish scholar Maimonides put much effort into reconciling Aristotle's cosmology with the Bible. Improving meteorology in the time of the Renaissance, however, would have meant not only outsmarting Aristotle but also defying the Pope.

The ancient meteorology Franklin dispelled was an intellectual mess, for Aristotle unfortunately was wrong about nearly everything. *Meteorologica* is divided into four books, one of which deals mostly with chemistry. In one bold stroke at the beginning of his treatise, Aristotle declares the atmosphere separate from the domain of the stars and the planets, thus disengaging meteorology, the study of the atmosphere, from astronomy, the study of outer space. This made some sense at the time simply because Aristotle considered the heavens immutable, made of alien substances. The atmosphere, meanwhile, was volatile and transient.

Aristotle borrows from some philosophers and repudiates others to piece together the first comprehensive system of weather knowledge. The system rests on two decisive assumptions borrowed from other eminent Greek thinkers. First, Aristotle adopts Eudoxus' tenet that the Earth is at the core of a universe of concentric spheres. The moon's orbit is the boundary sphere between the atmosphere and the heavens. A second idea, probably from Empedocles of Argigentum, a Sicilian of the previous century, was the essence of the four elements.

The four elements—fire, air, water, and earth—were not the (somewhat) irreducible substances that populate today's periodic table. They were qualities, or forms of existence, that changed into one another or mixed to create new qualities. The scheme proved richly nuanced and unfortunately perfectly suited for speculation about the sky. Sunshine triggered a series of transformations between elements that brought about weather. Ironically, this concept was close to our understanding of the weather today, for we know that the sun drives the cycling of moisture and heat and air around our globe. The varying angles of sunlight force the air to distribute unequal solar heating. The unequal heating of different ground cover—rock, trees, water, sand—also creates wind.

Aristotle's world worked differently, even if he took similar factors into account. For instance, the heat of the sun's rays incited the Earth to emit a rising exhalation, a hot and dry substance prefiguring fire. Similarly, the oceans—the layer of the world overlying the solid Earth—emitted hot, moist vapors. These exhalations and vapors were lighter than their parent elements and sought their proper, higher level in the universe. (The rigid structure of elemental levels in Aristotle's system helped explain the special chaos of an ocean storm, since winds diving into the watery sphere had left their proper level.) Storms took on specific character based on the vapor and exhalations mixed within. This mix was influenced by the daily cycle of the sun, the angle of the solar rays, and the proportion of land or water heated by sunshine. A thunderstorm with little or no rain might have an overabundance of dry, hot exhalation, while a deluge might mean an overabundance of vapors.

Vapors condensed in the frigid cold of the middle of the atmosphere. At first glance, this drop in temperature with height seems to correspond to the temperature profile we now know exists in the lowest 40,000 or 50,000 feet of the atmosphere, the stormy region called the troposphere. Aristotle figured correctly that the heat for this part of the atmosphere came from the Earth's surface and concluded that air would be cooler higher up. At even greater altitudes in Aristotle's system, the atmosphere heats up again. Similarly, today we are aware that temperatures steady in the tropopause and then increase again in the next higher level, the stratosphere. Far higher, at altitudes exceeding 100 miles, we know temperatures are hundreds or even thousands of degrees Fahrenheit (though it actually wouldn't feel hot in such thin air).

This comparison is deceptive, however. Aristotle figured that the sphere of air reached only the highest mountains. Beyond that lay the domain of fire, which he believed heated the top of the sphere of air. The idea lived for millennia. Even Johannes Kepler, the 17th-century astronomer who proved the elliptical orbits of the planets, wrote, "What we call air is not so high as to surpass all mountain peaks." He thought Mt. Olympus reached the top of the atmosphere. Christopher Columbus (according to his nephew Ferdinand) caused a stir by reporting that some of the peaks of the New World surpassed the height of the highest clouds.

The Aristotelian atmospheric temperature profile had implications as significant as the vertical temperature charts weather forecasters routinely construct today from balloon soundings. Modern meteorologists look for greater-than-normal decreases in temperature with height—an instability that encourages rising air, the stock in trade of thunderstorms. Without thermometers or balloons, Aristotle and his followers had no way to measure this lapse rate, but they believed that temperature contrasts encouraged cloud formation and caused rain or hail (among the fleeting phenomena called "meteors" in the sphere of air). Aristotle believed hot and cold repelled, so that small areas of hot vapors withdrew into clumps in an onslaught of cold, yielding condensation.

This is the kind of Aristotelian logic that turned nature on its head: we know now that most clouds signify expansion, not compression. Water condenses in air that cools by rising and expanding into lower pressure at higher altitude. Rising currents—at times violently rising updrafts—are a defining characteristic of severe thunderstorms. In Aristotelian meteorology, moist vapors rose simply to seek their natural level midway between water and fire. Rising or falling air—the "pockets" that jostle us in our airplanes—basically didn't figure into Aristotle's atmosphere. Any vapor rising too high would quickly explode into fire.

Aristotle had the sense to say that clouds were droplets of water that were small enough to rest on air. Small objects can float on the surface of water in a similar fashion, he argued (not realizing that, say, a waterbug skates on the surface of a pond in part by taking advantage of a relatively powerful attraction, called surface tension, that binds water molecules). But because they disregarded upper-air motions, Aristotelians floundered when describing clouds. Anglicus Bartholomeus, born in 1220, was an English cleric who borrowed heavily from Aristotle to write a famous compendium of medieval knowledge. "A cloud is kindly hollow, with as many holes as a spoung," he wrote, "And therfore by entring and incomming of the Sunne beames, a cloude representeth and sheweth diverse formes & shapes and coulours." In the 1400s, William Caxton regarded the color of clouds and wrote that dark ones were thick and ready to rain. Precipitating clouds, he said, eventually become lighter and whiter, allowing them to rise and dry up by proximity to the sun. Leonardo da Vinci, whose paintings and notebooks evince great insights into the nature of sky light, borrowed from Aristotle to explain that cold compressed heat and moisture tightly, forming ball-shaped clumps of cloud.

The hot exhalations eventually escaped those clumps, according to Aristotle, leaving chilled vapors to condense and return to the watery sphere. This was rain. Aristotle's description of the process clearly influenced thinkers as late as Francis Bacon, whose pursuit of verifiable evidence transformed the intellectual landscape in the 17th century. Bacon wrote: "Windes do contract themselves into rain" when they encounter mountains or opposing air currents.

To explain hail, however, Aristotle worked himself into a theoretical bind. He didn't know that vapor could freeze directly into ice, a process that we now know is critical in snow and hail formation. Aristotle knew that large hailstones are very big masses of water, but he had already professed that cloud water could float only if it remained in tiny droplets. The founding meteorologist worked out this dilemma by noting that the largest raindrops fall on warm days with violent rains. He concluded that hailstones form in warm clouds that trap cold air at low altitudes, triggering sudden, violent freezing (that hot-versus-cold reaction, again!). His proof: large hailstones are spiky, not smooth, and so have a configuration that would break apart if they had to fall far. As a result, Aristotle completely missed the connection between severe thunderstorms and the violent updrafts that can suspend heavy balls of ice long enough for them to grow layer by layer. Such updrafts also routinely launch clouds, and thus ice, five times higher than the gods at play on Mt. Olympus.

Aristotle established meteorology almost exclusively with shaky analogies and yet had the intellectual virtuosity to make room for all sorts of bizarre observations. Future scholars could explain nearly anything in Aristotelian terms. William Fulke used hot-and-cold contrasts aloft to explain all sorts of effects in his 1563 pamphlet, "A Goodly Gallery with a Most Pleasant Prospect, into the Garden of Naturall Contemplation, to Beholde the Naturall Causes of All Kind of Meteors." The young priest, who later became vice chancellor of Cambridge University, wrote that vapors congeal when south winds bring warm air close to cold winds, which isn't so far from the modern idea of cold and warm fronts. But Fulke's "Goodly Gallery" showed one of the important flaws of Aristotle's method. Fulke categorized meteors by cause; hence comets, thunder and lightning, winds, and earthquakes were lumped together as meteors of dry exhalations, while rain, snow, sleet, and hail were treated separately as meteors of moist vapors. This division unfortunately meant that the thunderstorm as we see it today didn't really exist in Aristotelian eyes. It would take a much later generation to show the connection between a hailstorm and a rainstorm and a lightning storm.

But an even less worthy meteorologist than Fulke was Du Bartas, a Renaissance author who subscribed to the popular idea that new life arose spontaneously from the environment. In some versions of this belief, air spawned life; in others, water. Some people believed that mud generated frogs, and Du Bartas said that when frogs spilled out of the sky, it proved that clouds contained all the elements essential to breed life: in the "active windes sweeping this dustie flat,/Sometimes in th'aire some fruitfull dust doo heape,/Whence these new-formed ugly creatures leape."

Storms, of course, mean not only precipitation but also damaging winds. Several philosophers before and after Aristotle realized that wind is simply the movement of air. Democritus, the Greek whose ideas on essential particles presaged our own atomic theories, claimed that densely packed air atoms caused wind, while loosely packed air atoms meant calm. None of these ideas swayed Aristotle, however. He thought that exhalations from the Earth accumulated little by little, like the streams from mountains, until a general rising flow gathered like a mighty river of fiery substance above the Earth. The movement of heaven and Earth, not a motion of air, made people perceive this exhalation as a horizontal wind.

Thus, just as rising air had no place in Aristotle's meteorology, neither did the horizontal wind. A storm without either, of course, is not much of a storm. Taking the error further, upper-level winds also did not exist in this cosmology. The jet-stream-enhanced dynamics of storms in today's meteorology were unthinkable in Aristotle's terms. When the ancient Greeks looked up at fall streaks spreading out from under high-level, wispy cirrus, they didn't imagine that the resulting shapes revealed the movement of air high above the Earth. They thought air aloft was relatively still; any disturbance high in the air would surely explode on contact with the volatile sphere above.

Enough scholars respectfully disagreed with Aristotle on the nature of wind to keep the notion of moving air alive into Franklin's time. Fulke was a good intellectual politician, striking a middle ground with a three-tiered classification of winds: on the broadest scale were general motions of the atmosphere, propelled by Aristotle's exhalations; in the regional rank were "particular wyndes" (like France's mistral) that escape from caves and holes in the ground, a force described by the Roman natural philosopher Pliny; the gentlest winds were moving air, as explained by Hippocrates. As for the strong winds that often precede sudden showers, Fulke explained (after Pliny) that a powerful wind was a "thycke Exhalation violently moved out of a cloude," causing perforations in the wall of the cloud through which water spilled to the ground. Pliny went further, claiming that whirlwinds started as winds trapped in a cloud, spinning around trying to get out. His remedy for the situation: throw vinegar at the vortex.

No matter what they thought caused wind, most observers related the winds to changing weather and approaching storms. Bartholomeus saw that north winds brought dry weather and south winds wet weather to England and concluded that these winds take on the characteristics of their origins: the frozen North and the southern seas. Had Bartholomeus extended this idea to the origins of air masses, the large, uniform high-pressure areas that dominate today's weather maps, he would be hailed a meteorological hero today.

Dense exhalations bursting out of clouds formed a semblance of a unified storm theory for Aristotle's heirs. The exhalations burst cacophonously into flame as they tore through cloud walls, creating lightning and thunder (far cry from the Epicurean notion that falling stars made lightning when doused by clouds in a final, sizzling blaze of glory). At least theorists associated lightning with thunder and knew that the two originated simultaneously.

Like other Renaissance thinkers, Fulke believed that damaging lightning was a projectile called "thunderstone," the most intense compression of the fiery exhalation propelled from the clouds. Often called a thunderbolt, the cooked stone could pierce armor and buildings. Similarly, Scandinavians believed that Thor's thunderbolts were solid pellets. People sometimes picked up meteorite fragments and identified them as thunderstones. Of course, there is a solid lightning residue—fulgurite, which is typically found in long, crooked strands buried in sand. It's a delicate, glassy, flaky mineral made when lightning fuses soil along watery pathways.

This brittle finger of heat-fused earth—fulgurite—is a nine-inch section of a world-record, 16-foot vein excavated in Florida in 1996. Fulgurite forms instantly from soil when lightning races helter-skelter toward underground water, but the Norse considered the mineral fragments of the spears hurled from the clouds by the god Thor. Photo courtesy of the University of Florida Lightning Research Laboratory.

Lightning wreaked havoc with devout and inquisitive minds for many years. Nothing in Aristotle's meteorology could prevent lightning from being the flash point for some of the most persistent of all superstitions about storms. Pliny had helped cement the legitimacy of lightning superstitions by listing items he considered safe from the clouds. Among them were the laurel tree and the seal. In his book *The Profitable Arte of Gardening*, Thomas Hill added other safe objects, including hippopotamus hide, owls (when their wings were spread), and speckled toads. The pillars of buildings were safe, too; long cylinders supposedly defied the jaggedness of lightning.

Perhaps to help children during stormy nights, Plutarch passed on the notion that people are immune from lightning while they sleep. Of course, more sensibly, parents must sometimes remind children that thunder is harmless. This truth is an old one, formerly stretched into a belief that thunder is actually beneficial. Plutarch, for instance, probably first published the legend that thunder helps mushrooms grow. Folk wisdom held that thunder helps baby swans break free of their eggs. And Thomas Hill, in another of his handy books, claimed that thunder purges the air of "evill vapours ..., yea the pestilence and other contagiousnese ...," underscoring the link between meteorology and medicine forged by Hippocrates in the fifth century B.C.

Many folk beliefs are ultimately reassuring and harmless. Probably only a few people died because they hid under laurel trees during thunderstorms, and it's a good bet that still fewer were struck by lightning while huddling under hippopotamus hide. But some beliefs about lightning were quite serious, for they were matters of faith. A crucial moment in the history of theology and meteorology may have been the battle Marcus Aurelius won near the Danube River more than 1500 years before Franklin's birth. A Hungarian tribe known as the Quadi surrounded the Roman troops and cut off their water supply. As the Quadi waited for the Romans' inevitable surrender, a rain began to fall. The Quadi attacked, realizing their siege was now futile, but just then the storm turned fierce. Wind and large hail blinded the Quadi and battered their bodies. Lightning from all sides struck them down in heaps as the Romans watched from safety.

Various versions of the story circulated in the Empire. In one, Arnuphis, an Egyptian sorcerer, came to Marcus Aurelius' aid, enlisting his gods to the Empire's cause. Roman pantheists, however, said the storm was a result of entreaties to Jupiter (who, being the supreme god, like the Greeks' Zeus, brandished lightning bolts as a weapon). Appropriately, believers celebrated the deliverance by building the Antonine Column in Rome, which shows Jove making the Quadi's meteorological misery. But the version of the story with lasting import in Europe was the one told by early Christians. They claimed that their prayers had brought the storm that delivered the Empire, and St. Gregory of Nyssa and St. Jerome preached this miracle. Combined with Biblical evidence of the Deity's control of weather, the story helped confirm the supremacy of Christianity far and wide.

Christian leaders by and large saw lightning as a moral force. A booklet published in Zurich in 1731, *Spiritual Thunder and Storm Booklet*, contained more than 300 pages of Protestant prayers and hymns for stormy weather. In the preface, the pastor at Nuremberg's St. Sebald church claims God uses storms in four ways: to show anger, to demonstrate power, to drive sinners to repentance, and to give people a taste of the Last Judgment. In a fable published in a book popular among nuns in the 13th to 16th centuries, lightning strikes a priest who is ringing the church bell. His clothes are ripped from his body, and his corpse is charred—a sign that the priest has broken his vow of chastity. Commented the Bishop of Voltoraria in the 1600s, "It is not to be doubted that, of all instruments of God's vengeance, the thunderbolt is the chief."

Theological explanations for lightning proved particularly insidious. Some say it was during Charlemagne's reign that people first rang church bells during storms to ward off lightning. Charlemagne, realizing the foolishness of the practice, apparently tried to outlaw the baptism of bells, a rite that made the sound of bells a lightning antidote. The ringing, however, continued. Baptismal instructions published in England in the 1500s advocated first washing the bell, tracing the sign of the Cross on it with oil, and then praying that the bells would dispel "hayle, thondryng, lightening, wyndes, and tempestes, and all untemperate weathers...." Many church bells bore inscriptions attesting to their powers: "On the devil my spite I'll vent/And God helping, bad weather prevent" a Swiss bell proclaimed (in German). Another boasted, "The sound of this bell vanquishes tempests, repels demons, and summons men," and another, "I praise God, put to flight the clouds, affright the demons, and call the people." Thomas Aquinas wrote, "It is a dogma of faith that the demons can produce wind, storms, and rain of fire from heaven."

The belief was too potent to be controlled. In the Middle Ages and Renaissance, thousands of people, mostly women and children, were forced to confess to inciting hailstorms or deluges with sorcery. Many were tortured and burned. Skeptics warily pointed out that storms continued unabated, but Luther himself preached that if one were to throw a stone into a pond near where he grew up, devils would be released by the water and cause a storm. Aristotle did not deal in demons. Nonetheless, he was by then the official natural philosopher of the Church, having been admired by St. Augustine and Bede the Venerable. A woodcut in a 1519 edition of his *Physics*, published in Augsburg, shows devils at play in the clouds, controlling the weather.

In a 1718 storm over Brittany, two bell ringers died and 24 different churches were struck by lightning. Amazingly, not one church tower that remained silent in the region was struck during the storm. One medieval author counted more than 100 bell ringers killed in three decades. Despite mounting evidence that climbing a bell tower during a storm is practically suicidal, the practice continued well into the 19th century in some regions. The Enlightenment had brought many changes, but popular belief was more powerful than reason when it came to thunderstorms. Even Descartes and Bacon endorsed the dangerous bell ringing, arguing that a subtle shock wave of air from a bell might disperse lightning. In Newton's time, the rector of Clementine College in Rome counseled that a ringing bell might "disturb and agitate the air, and by agitation [it] disperses the hot exhalations and dispels the thunder." But he added that, more certainly, the bell achieves its purpose by the "moral effect" of calling the faithful to prayer.

These signs of a rapprochement between science and religion in the Christian world helped pave the way for Franklin's discovery. The Puritan tradition of New England in some ways ushered in the new age. Puritan clerics believed that natural philosophy enhanced their ability to appreciate the mysteries of the world. In England, they applied Newton's laws to the Biblical Creation story, the Flood, and other events. Increase Mather and his son, Cotton, both preachers in colonial Boston, continued this tradition in the colonies. The Mathers loomed large in New England affairs for more than half a century, beginning in the late 1600s. At times, their Puritan ethic made them a seemingly Janus-faced force in the pursuit of knowledge. On the one hand you have Increase, the fiery orator of "The Voice of God in Stormy Winds," published in 1704: "When there are great tempests, the angels oftentimes have a hand there in,… yea, and sometimes evil and angels." To the elder Mather, God spoke through the whirlwind and punished the blasphemous with lightning. On the other hand you have Increase, the president of Harvard for 16 years, who had the integrity and curiosity to examine the bright comet of 1682 with the university's telescope and was impressed enough to help found a society in Boston to advance natural philosophy. And he obsessively investigated severe storms in the area, looking for signs of divine will as much as for insight into the workings of nature. Cotton Mather was a rare American Fellow of the Royal Society in London. His bold, life-saving experiment in inoculation during a smallpox epidemic in Boston in 1721 justified the prestigious honor. The Mathers together were also a formidable force against astrology. Increase commented wryly in a 1683 treatise on comets, "If men did with understanding read the Scriptures more, they would mind Judicial Astrologers less." As bleak as the state of storm theory was in the 18th century, the state of storm forecasting was even bleaker. In the absence of anything better, forecasters relied on astronomy and astrology for their predictions.

The Mathers were only a beginning. Against a tide of superstition, religion, and ages of scientific neglect, Benjamin Franklin righted the ship of storm science and set a new course toward rapid advance. Experiments with rubbing spheres and tubes of glass had shown him that pointed objects don't produce sparks nearly as easily as round objects. In his new theory of positive and negative charge, Franklin explained that spheres uniformly distribute charge on their surface, while pointed objects pool charge at their tips. Then he put this finding together with another observation. In his notes of November 7, 1749, he took stock of the similarities between lightning and electricity—the light, the color, the crookedness, the speed, the ability to pass through water or metal, the noise, the damage, the sulfurous smell. "The electrical fluid is attracted by points," he observed. "We do not know whether this property is in lightning." So Franklin proposed an experiment—the first experiment of his own design—to show that clouds and lightning were electrical. He proposed erecting a sentry box on the roof of a tall building. Inside would be a Leyden jar attached to a pointed rod jutting up into the sky. The jar, insulated from the ground, would receive charge from the cloud and store it. Franklin expected a slow discharge of electricity, not a lightning strike per se. Franklin wrote about the sentry box to Collinson, his correspondent in London. The idea was published there in 1751 and soon translated into French as well.

Meanwhile, Franklin was waiting for a suitable tower with which to conduct his experiment. By June of 1752, he gave up waiting for the construction of the new spire at Christ Church in Philadelphia. One day that month (as near as historians can tell), a thunderstorm loomed over the Philadelphia area. Franklin and his son, James, secretly fashioned a kite out of a large handkerchief and quietly set out for a nearby field where no one could watch what they anticipated might be a disappointing experiment. From the relative safety of a shed, they lofted the kite by means of hemp string with a brass key tied to it. They expected the hemp to moisten and conduct charge to the key. To play it (relatively) safe, Franklin didn't hold the hemp string itself, but instead controlled it with a dry silk cord. A vanguard of cloud passed overhead and nothing seemed to happen. But then the elder Franklin noticed strands of hemp standing up like the quills on the back of a startled porcupine. He put his knuckle up to the key, and even though the string was still mostly dry, he drew a spark. Later during the storm, as the string moistened, even more electricity flew between Franklin and the key. For good measure, the thorough scientist found he was able to charge his Leyden jar with the apparatus.

Franklin didn't wait long to suggest the natural corollary to his experiment. The next annual edition of *Poor Richard's Almanack* described a pointed rod that would drain charge from clouds overhead before light

Engineers take Franklin's lightning rod concept a step further, testing power line durability by enticing the clouds to erupt with lightning, like this wind-feathered banner of electricity. Courtesy of the University of Florida Lightning Research Laboratory.

ning could strike. Franklin also studied buildings struck by lightning in Philadelphia and knew that in the event the rod was struck by lightning, the charge would be conducted harmlessly to the ground, saving the building. In 1752 the first lightning rods were introduced with success in Philadelphia.

Franklin didn't realize that the kite experiment was actually moot. A month earlier, the King of France had read with great interest of Franklin's sentry box proposal and had encouraged his top scientists to try it. On May 10, 1752, in Marly-la-Ville, Dalibard set up a 40-foot metal rod in a garden. With a local priest among the witnesses gathered to attest to the results, the first lightning rod drew a spark. Within the year, other Frenchmen repeated the experiment with taller rods, drawing charge not only from ordinary clouds but also from clear air. The success of the lightning rod was repeated in Belgium, Germany, Ireland, and England soon after: Franklin's fame spread like wildfire in Europe before he even knew the experiment worked. And contrary to popular belief, and Franklin's own worries, the experiment proved relatively safe, though the German scientist Georg Wilhelm Richmann was killed by lightning while demonstrating the procedure in St. Petersburg, Russia.

Plaudits for Franklin's achievement came fast. The Royal Society had not yet even made Franklin a member, yet it took the unprecedented step of awarding an outsider its Copley medal, the equivalent of a Nobel Prize today. The equivalent academy in Paris honored Franklin by making him its only foreign member, a status that gave him a distinct advantage later as the rebellious colonies' representative in France. In America, a Baptist minister and scientific lecturer named Ebenezer Kinnersley began demonstrating how Franklin's lightning rod could protect wooden models of houses. With a spark from the static generator, the unprotected model would burst into flame while the protected model remained unharmed.

Kinnersley and other proponents of the lightning rod ran into resistance from some religious leaders who considered it presumptuous that mankind should erect a shield against God's hand. In Charleston, South Carolina, a newspaper got around the problem by urging the construction of lightning rods "to the glory of God." Church officials who didn't heed Franklin's invention continued to pay a price. San Marco in Venice had been damaged or destroyed by lightning seven times before 1752. Church officials nonetheless balked at erecting Franklin's device until two more thunderstorms, in 1761 and 1762 damaged the tower again. A rod was erected atop the church in 1766, and it has weathered lightning safely ever since.

Bostonians installed so many lightning rods that the Reverend Thomas Prince blamed them for the earthquake that rocked New England in 1755. Prince misunderstood Franklin's discovery and assumed that the electrical charges of clouds acted not alone but in concert with "sulphureous, nitrous, mineral, watery, and airy Substances ... as a principal Means of exciting them in Action." The whole connection between lightning and earthquakes was ancient, dating back to the idea that an excitation of exhalations trapped underground could cause the ground to shake. Prince thought the lightning rods had drawn too much charge into the Boston terrain, not realizing that charge in the earth is distributed relatively uniformly, as it would be on any sphere. "O! there is no getting out of the mighty Hand of God," he bellowed. "If we think to avoid it in the Air, we cannot in the Earth: Yea, it may grow more fatal."

But Prince's plea was a late gasp for a dying culture of Aristotelian understanding. Storm science had moved on to another phase, and it was now dangerous to ignore not the mighty hand of God but the powerful ideas of Franklin. King George III was one of the last to risk opposing the sage colonial. The King blamed Franklin's influence for the American Revolution, and then, in 1777, he blamed him for another British catastrophe. In that year His Highness' ammunition storehouse at Purfleet suffered damage from lightning. It had been protected by pointed lightning rods of Franklin's design, and the King ordered new, blunt rods to replace them at Purfleet and the palace in London. Sir John Pringle, president of the Royal Society, defended Franklin's observation that pointed objects draw charges most easily. The good knight was by virtue of his office the chief science adviser of the realm and wrote to his King, "I cannot reverse the laws and operations of nature." King George, in a characteristic gesture of pride before prudence, promptly dismissed Pringle. He had already proven he would only learn from Americans the hard way. This incident turned out to be one of the mad King's few victories against Franklin, for blunt rods turned out to be, in some cases, preferable to pointed ones—a minor negotiation settling a war already won against superstition.

CHAPTER THREE

Storms on the Move

Rife with the usual follies of battle and diplomacy, the Crimean War of 1854 was an infamously misguided moment in history. For starters, the war was based on idle bluster: Tsar Nicholas I used hollow religious pretexts to open a port for his Navy on the Black Sea. France's new emperor, Louis Napoleon, eager to crown his regime with easy glory, pounced on the opportunity to help the British punish the Russians. Incompetence also reigned on the field. The Allied troops landed on the Crimean peninsula in the middle of unseasonable 90-degree heat and immediately shed the insufferable wool blankets and coats they would desperately need later. Quickly, the Allies took a commanding position near the Russian naval port of Sebastopol. Yet on October 25, a poorly armed phalanx of British cavalry suffered a hopeless massacre in the teeth of Russian artillery fire, a bungled moment of bravado immortalized in Tennyson's poem "Charge of the Light Brigade." The Allies finally disposed of the Russian army at the Battle of Inkerman on November 8, 1854, but then squandered their advantage. Instead of attacking Sebastopol, they chose to lay siege on the port.

With its legacy of tragic negligence, it is only fitting that the crowning mistake of the Crimean campaign was a costly disregard for the weather. The troops settled in at Balaclava, a port a few miles south of Sebastopol. A week into the siege the favorable Indian summer came to an abrupt end. A secondary low-pressure center spun off a routine storm in Europe; maritime moisture over the Black Sea pumped it to extreme intensity. Rain began falling on November 14, and the next morning an unexpected gale roared in from the southeast. The flimsy troop tents blew away. Winds shifted to the west, topping 70 m.p.h. as snow began to fall. The finest ship in the French fleet, the *Henri IV*, sank in the rough seas. Worst of all, supply ships in the harbor were dashed to thousands of pieces. On one 7,000-ton steamer alone, 40,000 badly needed coats and other winter necessities sank to the bottom of the harbor.

As the wind relented, a wave of cold air settled in and temperatures dropped drastically. Lighting a fire was impossible in the breezy, wet conditions. The men's teeth chattered; their horses froze. The few healthy soldiers carried provisions salvaged from the harbor to higher ground. They knew if they took their boots off their frozen, swollen feet, they would never be able to get them back on. Within two weeks, more than 8,000 men were gravely ill, most afflicted with cholera. The Army's medicines lay underwater with the rest of its supplies.

Soldiers who were evacuated to the ill-prepared British hospital in Scutari, on the Turkish coast, endured appalling conditions. The wounded lay on floors, and the hungry ate uncooked meat. Doctors were too busy to wash their operating tables. By early 1855, soldiers were dying at a rate of 3,000 a month, from diseases festering at the hospital as much as from the cholera epidemic. An unheralded nurse, Florence Nightingale, quickly rounded up women in England and raced to Turkey to restore sanitation (and sanity) to the hospital. Not until February, four months after the storm, did ships from the Admiralty relieve the starving men in Balaclava.

Public outrage against the war pinned both the French and British governments against their political backs. Marshal Vaillant, France's minister of war, didn't expect decisive answers from the French Academy of Sciences' official investigation into the storm. So in January 1855 he sidestepped that august body and asked Urbain Jean Joseph Le Verrier to look into the matter independently. Le Verrier was head of the Imperial Observatory in Paris, which recorded weather data, but he was more famous as an astronomer. In 1846, Le Verrier had scratched a few calculations on a pad of paper and used them to discover a new planet, Neptune.

Le Verrier at once wrote to observatories in Europe asking for their weather records for November 11–16, 1854. More than 200 of his colleagues responded. By February 1855 Le Verrier could see that the storm had been evident long before it struck Crimea. Clearly, if the Allies had thought to gather weather observations systematically, they would have seen the storm coming. The information could have saved thousands of lives. On February 16, 1855, Le Verrier urged Louis Napoleon to approve a plan for a storm-warning service. The loss of 400 more lives that month when a troop transport ship sank in a storm added urgency to the situation. The French leader, pressed to save face after Balaclava, approved Le Verrier's plan within a day. Three days later Le Verrier began to collect weather data on a regular basis from around Europe; within six years, every major European city was part of the weather-reporting service.

The British, too, responded to the crisis with a meteorological answer. Parliament established a system for collecting weather data from ships' logs. A month after the Balaclava fiasco, Captain Robert FitzRoy was put in charge of England's new cooperative weather effort. Just to cover their tracks of neglect during the war, the Admiralty backdated FitzRoy's orders to August 1854, before the Crimean expedition. Shortly afterward, FitzRoy moved to the new Meteorological Department, under the Board of Trade, to collect and publish observations of current weather around the nation.

Le Verrier and FitzRoy's weather reports represent the great progress in storm science up to the middle of the 19th century. In the 100 years since Franklin had tamed lightning with his kite experiment, meteorologists had begun searching for the laws behind the weather. They discovered that storms had shape and history. They confirmed that storms stretched beyond the horizon, lasted for days, and traveled many miles. To study cyclones that spread 1,000 miles across oceans and continents, scientists made the most of the emerging technologies in the 19th century, just as they would do in our own century. They also pioneered a unique cooperative approach suitable for a research problem of this size.

As the science historian James Fleming has shown, meteorologists of the 19th century perfected scientific teamwork out of necessity. The discoveries of the era made clear that meteorologists could not do their research alone, relying solely on their own observations. They learned to work together, and even to this day, thousands of volunteers create networks of observers to measure the vast atmosphere. FitzRoy's new job, for instance, was a British response to an international effort started a few years before by Matthew Maury of the U.S. Naval Observatory in Washington. Maury had gathered scientists from around the world to agree on the free exchange of meteorological data to complete worldwide charts of winds and currents. Maury's principle remains at the core of meteorology more than a century later. Even during the Cold War, international distribution of satellite observations and upper-air data was practically unfettered.

The push toward systematic cooperation on storm science began with a critical discovery by—who else?—Benjamin Franklin. Not only was the lightning rod the biggest nail in the coffin of Aristotelian meteorology, but, on a cloudy, windy evening in 1743, Franklin also launched the transformation of storm science in the 19th century. Instead of his kite, he used the mail. On Friday, October 21, of that year, Franklin had little interest in storm watching. Instead, he was hoping to catch a lunar eclipse at 9 o'clock that evening. To his disappointment, not long after sunset, a northeast wind swept in thick clouds overhead, blocking the eclipse. It was one of the already infamous nor'easters that badger the East Coast much of the

The late-19th-century Cuban priest Benito Viñes earned the nickname Father Hurricane for his storm forecasts. This wheeled cloud chart, based on his principles, in turn based on William Redfield's storm theory, helped the user locate the eye of an approaching hurricane. Photo courtesy of Patrick Hughes.

year, especially during fall and early spring. Franklin's disappointment turned to curiosity, however, when the mail arrived. A Boston newspaper reported that the eclipse had dazzled residents, only to be followed by a nor'easter around midnight.

"This puzzled me ...," Franklin later explained, "being a N.E. storm, I imagined it must have begun rather sooner in places farther to the northeastward than it did at Philadelphia." Sensing a discovery at hand, Franklin wrote to correspondents up and down the East Coast for their observations of this and other storms. He soon realized that the direction of the wind had little to do with a very obvious southwest to northeast progress of most storms in the area. In the case of the eclipse, Franklin calculated the nor'easter had moved the 400 miles toward Boston at a rate of 100 m.p.h.—a huge overestimate. Nonetheless, Franklin had documented irrefutably the motion of a storm. Large storms could never again be mistaken for local phenomena. If storms moved, then they could be tracked. If scientists could figure out why they moved or where they tended to move, then they could forecast storms. Franklin's successors would soon follow these invaluable leads.

But one other critical question remained, which Franklin uncharacteristically mishandled. His nor'easter study revealed that at least some winds within storms blow contrary to the motion of the whole. This had been documented long before anyone knew storms moved. The first observation of the whirlwindlike nature of large storms was recorded by the infamous pirate Will Dampier of England. Dampier made many first scientific observations on his three voyages around the world in the late 17th century. One could argue that he used privateering merely as a way to pay for the chance to see the world. Fortunately, Dampier kept careful records and an open mind. In 1699 he took time out from marauding and pillaging to make global wind charts that in many ways were still considered state of the art at the beginning of our own century. On one of his trips Dampier ran into a typhoon in the China Sea. From observing the shifting winds that threatened to tear apart his ship, Dampier realized that the tropical storm was a giant counterclockwise circle of wind. He related the typhoon to its American cousin, the hurricane, which had become quite familiar to Spanish galleon commanders in the 1500s.

Franklin was aware of whirlwinds on a smaller scale—tornadoes, waterspouts, and other vortices—but he did not deem them relevant to American nor'easters. Though he was right about this, the whirlwind turned out to be a far better approximation of a nor'easter, or any midlatitude cyclone, than Franklin's straight-wind model. Franklin said that tropical heat thins, or "rarefies," the air over the Gulf of Mexico. The tropical air rises, and in response dense air far away moves to fill in the Gulf atmosphere. Air from the East Coast travels southwestward because it is channeled by the Appalachian Mountains. Franklin said the rising tropical air was like a gate lifted at the end of a sluice. In a sluice, the water nearest the gate flows first, followed in turn by water farther and farther away. Similarly, the southernmost air along the East Coast responds to the heating in the Gulf first, and in turn air farther and farther up the coastline begins flowing southwestward.

William Redfield, a native of Middletown, Connecticut, set Franklin straight almost 100 years later. Franklin would have been proud of this practical, self-made talent. Redfield's father, a sailor, died when William was 13, forcing the youngster to leave school and become a saddlemaker's apprentice. Nonetheless, Redfield retained an insatiable appetite for learning, and a kindly local doctor lent him countless books. The teenager couldn't afford a good lamp, so he read by the fireplace after work. The circumstances were no impediment, and Redfield surprised his benefactor by returning the books quickly and selecting more and more technical volumes. In 1810 Redfield finished his apprenticeship and took the opportunity to visit his mother, who had remarried and moved to Ohio. He made the 700-mile trip on foot, each evening taking notes on what he saw. The next year he walked back home by a different route, filling his journals with new insights. Redfield became a small-town mechanic in Connecticut first but later moved on to be an engineer and manager of a busy river barge service based in New York City. The observations he took on his trip to Ohio were eventually used to establish a railway route across the Appalachians.

In 1821 Redfield again spent long hours walking the countryside and forests of Connecticut, now as a traveling salesman based in Cromwell, near Middletown. His walks were again fruitful. On September 3 of that year a great hurricane ravaged the Eastern Seaboard. Before dawn it swiped the North Carolina coast, killing several people on Currituck Island and destroying at least a hundred houses. Next it wreaked havoc at the naval installation in Norfolk, Virginia, where the local newspaper described the experience as a "deafening roar of the storm, with the mingled crashing of windows and falling of chimneys—the rapid rise of the tide, threatening to inundate the town—the continuous cataracts of rain sweeping impetuously along darkening the expanse of vision, and apparently confounding the 'heaven, earth, and sea' in a general chaos." Late in the afternoon the storm raised tides on the Delmarva peninsula by 10 feet. Salt spray singed crops inland in New Jersey as the eye sped up the coast over the route of the current Garden State Parkway. Soon the devastating central winds were pounding New York City, making this the only major hurricane on record to march its eye through the city, according to weather historian David M. Ludlum. The Battery was inundated with crashing waves, and the city's wharves were destroyed. The eye came ashore at the site of today's Kennedy Airport and tossed ships onto the sands of Long Beach and Rockaway Beach before crossing Long Island Sound. Seventeen people died in the waters there when their ship sank. At 9 p.m., the storm rammed into the Connecticut shore and headed across New England and back over the Atlantic, where it finally dissipated.

Not long after the terrible storm, Redfield was back on his feet, embarking on a sales trip with his eldest son. Along the way he noted a curious pattern in the fallen trees. In Middletown nearly all the trees had been flattened toward the northwest by the southeast wind. But in northern Littlefield County, less than 70 miles away, trees pointed toward the southeast. The wind here, residents told Redfield, had blown from the northwest. Redfield began collecting observations of the storm, learning which way winds had headed in each community and finding out how they had changed suddenly from southeast to calm to northwest in some places.

At first Redfield did nothing with this information. Then, in 1831, Redfield struck up a conversation with a distinguished passenger on a boat from New York to New Haven. It was Denison Olmsted, a professor at Yale College. Olmsted was impressed when Redfield asked questions about the professor's recent article on hailstorms. More importantly, Redfield revealed his thoughts about storms and the evidence he had gathered 10 years earlier. Olmsted had never heard ideas quite like them. At Olmsted's insistence, Redfield published his findings in the *American Journal of Science* in 1831, in an article entitled "Remarks on the prevailing Storms of the Atlantic Coast, of the North American States."

In the prevailing classification of that time, the strongest storms were called hurricanes. The differences between hurricanes—which form in the tropics—and strong midlatitude storms were unknown. If anything, meteorologists classified storms along the American Atlantic coast by wind direction, as suggested by Franklin's linear storm theory. A nor'easter was cold with rain and maybe snow; a southeaster was warmer and less likely to bring wet weather. Redfield noted that experts labeled the 1821 hurricane a southeaster, but he proved that the storm was actually a "progressive whirlwind." Redfield rendered the directional classification of storms obsolete by showing that the air in storms moved in more than one direction at the same time—in 1821 he found opposite winds only 70 miles apart. Redfield compared storm circulation to water spinning in a basin stirred evenly around its edges. The water level in the basin lowers gradually toward the center. (By contrast, the surface of water draining out of a basin through a hole in the bottom lowers steeply near the hole.) Redfield claimed that the pressure in a storm decreases gradually as you move toward its center.

With this article Redfield the engineer, at age 42, began moonlighting as a meteorologist. He started to collect hurricane observations from ships' logs. He tirelessly documented Atlantic hurricanes. To plot the winds of an 1844 storm, he used some 164 different sources. Science suited Redfield well. By the time he was named first president of the American Association for the Advancement of Science in 1848, Redfield also was a renowned paleontologist and natural historian.

His Law of Storms won him lasting fame, even though his basic, one-whirlwind-fits-all approach did not survive close scrutiny later. Redfield believed that large midlatitude cyclones and hurricanes all follow a blue-print drawn, on a much smaller scale, by tornadoes and waterspouts. He admitted that weaker storms might not form regular spirals but insisted that the strongest storms are spirals of wind 1,000 miles across (accurate for extratropical storms, but more than double the size of most hurricanes). He believed that air gradually wraps inward until it circles a clear, central axis that may be vertical or inclined (true for hurricanes, but not for extratropical storms). As air nears the center it accelerates, ultimately reaching speeds between 100 and 300 m.p.h. (the hurricane to tornado range). As a whole, the whirlwind is capable of moving at anything from a walking pace to more than 40 m.p.h. Furthermore, Redfield said, the whirlwinds of the Atlantic (and the China Sea) typically follow recurving paths over the oceans. (We now know this only applies to tropical cyclones.) In the Atlantic, this path begins in "Equatorial regions" (actually, almost always more than 5 degrees from the Equator, satellites now show), stretches westward past the West Indies, then northwest and north, then northeast along the coast, and finally back out into the open Atlantic—a journey often exceeding 3,000 miles.

In Barbados, the English Colonel William Reid made an exhaustive study of a hurricane soon after the publication of Redfield's first paper. He confirmed the rotary theory and became one of Redfield's strongest sup-porters. Sailors around the world began to learn the usefulness of Red-field's Law of Storms. No less a figure than Commodore Matthew Perry endorsed its life-saving power, and the commandant of the famous voyage to Japan wrote an introduction to Redfield's "Essay on Pacific Cyclones."1 With Redfield's studies, Perry and other sailors now knew that in the northern hemisphere, at least, counterclockwise winds and decreasing pressure revealed where a tropical storm center lay. Redfield believed that the storm probably leans toward the direction in which it moves. The upper clouds—icy cirrus—race out ahead of the bottom of the storm, which is retarded by surface friction. Redfield's theory thus explained why, for centuries, sailors considered approaching cirrus to be an ominous portent.

In India, a British official named Henry Piddington realized that Bay of Bengal storms were also shaped like a coiled snake and coined the term "cyclones." Like Redfield, he used ships' logs for storm studies. One brig, the *Charles Heddles*, provided Piddington with startling evidence. The *Heddles* set sail from Mauritius on February 21, 1845, bound across the Indian Ocean for Madagascar. Not far from Mauritius the winds began to howl and waves crashed on deck, sending water down the hatches. The crew pumped the hold and prayed as their ship wandered at the mercy of the wind. After six days of overcast, the captain finally had a chance to read the sky. To his surprise the *Heddles* was practically where it had started the voyage. The winds had blown the ship in counterclockwise loops hundreds of miles across. Such experiences convinced Piddington that scientists would someday gather information by sailing into hurricanes, a vision of storm-hunting voyages realized 100 years later with airplanes.

Piddington's *Sailor's Horn Book* and Redfield's work helped mariners avoid the dangerous right front quadrant of the hurricane. Because the storm moves, not all parts of the cyclone have equally strong winds, even if the rotation speed is uniform. In the right front quadrant of a northern hemisphere tropical cyclone, the counterclockwise winds and the storm's overall motion coincide, increasing the local wind speed. Thus in a 100-m.p.h. whirl moving at 40 m.p.h., the winds blow at 140 m.p.h. in the front right but only 60 m.p.h. on the left. When a storm approached from the east, sailors would speed south, letting the wind carry them to the weaker front left quadrant. Of course this tactic had its dangers. Off the North Carolina coast, where hurricanes often move north, sailors heading west with the wind might dodge the right front quadrant but founder in the dangerous waters off Hatteras.

The spiral model worked so well that Redfield overgeneralized the rotary model to all strong storms. Nonetheless, even misapplied, the Law of Storms was a tremendous achievement. Coming on the heels of an era in

An 1819 cloud atlas shows the emerging understanding of how clouds work, with somewhat realistic cumulus towers in various stages of growth. More question-able, however, are the mixed altitudes of altocumulus bands, presumably pro-duced by waves of air over the mountains to the left. NOAA photo.

which storms and chaos were synonymous, Redfield's discovery of or-derly circulation in the severest storms was a revelation, a new maturity for a young science. In 1857, the year of Redfield's death, Olmsted wrote:

> In no department perhaps of the studies of nature [has] mankind been more surprised to find things governed by fixed laws than in the case of the winds. It is now rendered in the highest degree probable that every breeze that blows is a part of some great system of aerial circula-tion and helps to fulfill some grand design. Inconstant as the winds has long been a favorite expression to denote the absence of all uniformity or approach to fixed rules; but the researches of the meteorologists of our times, force on us the conclusion that winds, even in the violent forms of hurricanes and tornadoes, are governed by laws hardly less determinate than those which control the movements of the planets.

This feat—the discovery of order in weather—brought Redfield's work to the attention not only of mariners and commodores but also of scientists, including the astronomer Le Verrier. Astronomers had shown they could predict the motions of the celestial sky with thorough observa-tions and rigorous application of physical laws. Surely weather would respond to the same treatment. Redfield's storm model enabled Le Verrier to locate the Balaclava storm over the Mediterranean despite relatively sparse data. All the French scientist had to do was assume the storm rotated around low pressure, then fill in the gaps between observations. Indeed, the rotary storm theory quickly made maps the primary tool of meteorology. The theory made useful maps possible; at the same time, mapmaking was a way to test the theory.

Maps proved especially useful as meteorologists began to settle on atmospheric pressure as the signal variable in stormy weather. This idea by no means inspired unanimous assent, even at midcentury. But evidence had been mounting ever since the discovery of atmospheric pressure itself. The inventor of the barometer, Evangelista Torricelli, noted in the mid-17th century that pressure changes when weather changes. Not long after, his contemporary, the great Prussian experimentalist Otto von Guericke, took time from his day job as mayor of Magdeburg to note the loss of pressure during the passage of storms. He even claimed to be able to predict coming storms on the basis of his barometer. Guericke was also a pioneer for pointing out that air never freezes—only the water in the atmosphere freezes. This was before chemists isolated the constituent gases of the atmosphere—a fairly uniform proportion of nitrogen (78%), oxygen (21%), and traces of hydrogen, carbon dioxide, and other substances, and a tiny but highly significant dab of water vapor.

Redfield for one resisted the emphasis on pressure, which he considered a byproduct of wind. He ascribed storms to atmospheric tides that constantly shifted air across vast distances. Meanwhile, others began mapping pressure tendencies associated with storms—a thoroughly modern practice. Heinrich Brandes of Prussia drew the first of these modern meteorological maps in 1816. Brandes, a professor at the University of Breslau, was trying to link low pressure and storms and used pioneering observations made by the Palatine Meteorological Society of Mannheim. He noted the barometer readings and calculated how much they varied from normal (differences now called "departures"), then plotted the departures on maps. Each map showed simultaneous readings. He then drew contours to link equal departures. The contours clearly showed that storms reduce local pressure and that the region of this reduction follows the storm as it moves across Europe. None of this was terribly shocking— people had been diligently reading barometers for more than a century, and they now understood the movement of storms. But Brandes' map method also showed that weather was in constant turmoil. Even on a calm day his maps showed storms moving. Unfortunately, scientists ignored

Brandes' map technique, published in his 1820 book, perhaps because the storm he had depicted crossed the continent almost 40 years earlier: the Palatine Society data had been moldering in a library since the project ended in 1792.

Brandes' "synoptic" method—analyzing readings all taken at the same time—was too valuable to be ignored forever. His maps, with pressure contours surrounding the storm nexus, introduced an aesthetic now synonymous with storms in meteorology. In 1836, a young professor of mathematics at Western Reserve in Cleveland, Elias Loomis, also studied a storm synoptically, borrowing Brandes' method of a progressive series of maps of simultaneous data to analyze the motion and circulation of a storm. Then, for a severe storm in February 1842, Loomis made an innovation: instead of drawing lines of equal pressure departure from normal, Loomis used equal pressure contours. These "isobars" dramatically showed the intensity of the storm encircling the low-pressure center. The closer together the concentric rings of isobars, the steeper the drop of pressure near the storm's core, just as densely packed elevation contours mean steep slopes on a topographical map. The steeper the pressure gradient, the faster the wind. In addition, the movement of the circular bull's-eye from synoptic map to synoptic map showed the movement of the storm.

Synoptic maps certainly aided research, but not forecasting, at first. Brandes had shown that storms move faster than a horseman could deliver the data in 1816. Somehow, synoptic maps had to keep pace. Fortunately, along came the telegraph, one of the many technological breakthroughs that remade meteorology. As soon as telegraph wires stretched across the United States, people began sending messages about the weather. The telegraph operators were not trained meteorologists, but they knew that weather moved from west to east over the continental United States. So they began sending crude weather forecasts of their own across the wires to their colleagues farther east. Thus the high-pressure world of forecasting storms was born. Appropriately, the first scientist to develop the possibilities of the new medium was the American who invented the telegraph, Joseph Henry.

In many ways Henry's early life in upstate New York paralleled Redfield's boyhood. His father died when Henry was a boy, and his mother raised the family in poverty. Henry began working at age 10 in a shop 36 miles from home. At this time, Henry found a refuge in the village library. After several years apprenticed to a watchmaker, he showed interest in becoming an actor. But at 16, upon reading his first science book, Henry determined to head toward academia. Night school, tutoring, and teaching followed. Henry quickly rose from lab assistant to full professor at the Albany Institute, the equivalent of a high school academy. During summer vacations, Henry worked in the Albany labs replicating experiments and concocting demonstrations for his classes. In 1827, at the age of 30, he began improving electromagnets with coils of wires. Four years later, he surprised his students by stringing a mile of wire around the classroom. When electricity surged through the wire, it created a sudden magnetic field in one of Henry's specially made magnets of coiled, insulated wire. The magnet pulled a hammer on a pivot, clanging a bell. The schoolteacher's coil magnet had made history.

Following Benjamin Franklin's example, Henry refused to patent his work. "I did not then consider it compatible with the dignity of science to confine benefits which might be derived from it to the exclusive use of any individual," he later explained, adding, "In this I was perhaps too fastidious." Indeed, with a few improvements on Henry's basic design, Samuel F.B. Morse and others claimed fame and fortune with the telegraph. Henry, on the other hand, won acclaim as the greatest American scientist since Franklin. In 1846, after a stint as a professor at the College of New Jersey (later renamed Princeton University), he became the first secretary of a new national institution for scientific research in Washington, D.C.—the Smithsonian. At the expense of his own research career, Henry leapt at the opportunity to put American science on the map. Before the year was out, he proposed a network of weather observers, linked by telegraph to the Smithsonian, to help settle controversies in storm theory.

Each observer received calibrated instruments—a major advance over earlier networks—and instructions on when and how to use them. By 1860, 500 observers across the country volunteered for the project and took four observations a day, the first at sunrise and the last at 9 p.m. local time. Henry posted the readings as observers wired them to Washington. On the Mall, a giant map with color-coded disks showed storms cross the nation. Occasionally, Henry would venture a forecast for official Smithsonian events, in Washington, but mostly the project remained a test case. Unfortunately, because the observations were not simultaneous across time zones, they were useless for storm studies. (Today official observers around the globe record data at least at noon and midnight, Greenwich Mean Time, to facilitate the necessary synoptic maps and computer calculations.) If the information didn't serve Henry's primary objective, it later did advance climatological research. And the popular, real-time weather-reporting service, just a stone's throw from the Capitol building, deeply impressed lawmakers.

The Civil War shredded the national Smithsonian observing network and probably set back progress in American weather research by more than a decade. But the weather continued to be as stormy as ever. In 1868 and 1869, more than 3,000 ships ran aground or sank in Great Lakes storms. More than 500 people drowned. Americans clearly needed a storm-warning service like Henry's. In Europe, FitzRoy and Le Verrier had already established such networks. In 1869, Increase Lapham, a college professor and an astute Smithsonian weather observer in Wisconsin, urged his local congressman, Halbert Paine, to take note of forecasting successes in France and England and begin a similar service in the United States. Paine in turn persuaded his colleagues that the Army could tele-graph warnings to the public.

The military was a logical choice for the new weather service. For many years, beginning in 1814, Army medical officers had collected obser-vations for the Surgeon General in a study relating weather changes to epidemics. Now the Army's Signal Service was desperate to expand its role in the modern age. In February 1870, Congress approved the National Weather Service Act, handing weather duties to the Signal Service. By November 1, a network of 25 observer-sergeants were taking simultaneous readings (a great improvement over the initial Smithsonian timetables). General Albert Myer, head of the Signal Service and a former surgeon in the Army's medical meteorology network, asked Professor Lapham to distribute storm warnings for the Great Lakes region. On November 8, Lapham gave the first government storm warning in the United States as surface pressures were falling and winds increasing in Chicago and Mil-waukee and other Great Lakes cities. After only two weeks, though, the 60-year-old Lapham left the weather service in Chicago and headed back to Milwaukee. In January 1871 Myer found the right man for the job: a young astronomer named Cleveland Abbe.

Abbe had a passion for the sky and for science. His love of observation took him and several other men on the arduous journey from Washington, D.C., to South Dakota in 1868 to watch an eclipse. The Dakota natives were duly impressed by the white man's precise forecast for the celestial show—and probably even more impressed by his determination to see it. (In 1890, Abbe would travel to Africa to see another eclipse.) Born in 1838 in New York City, young Abbe proclaimed himself an evolutionist at age 9 after reading a book about the new theory. His greatest love as a boy was the night sky, however, so at age 14, Abbe enrolled in New York Free Academy, now City College, to study astronomy and math. Only at college, thanks to a new pair of glasses, was the severely nearsighted Abbe able to see the stars for the first time. After further studies in astronomy at Michigan and Harvard, Abbe went to Pulkova, Russia, to work at one of the world's best observatories. In 1868, eager to create a similar facility in the States, he jumped at the chance to head the Cincinnati Observatory, which then had the world's sixth-largest telescope.

The difficulty of studying astronomy in the bustling, smoky environs of a growing American city soon turned Abbe's attention to meteorology. Abbe now found himself at a place where local atmospheric conditions compromised his work. In addition, the observatory was dangerously short of money. Abbe knew he needed to make the institution a more integral part of the city's life. What better way than to combine his troubles and start a weather prediction service? Inspired by articles by the latest storm theorists, Abbe felt that the day had come for forecasting storms. "The science of meteorology is slowly advancing to that point at which it will begin to yield most valuable results to the general community," he said in his first address to his Cincinnati sponsors. "We can generally predict three days in advance any extended storm and six hours in advance any violent hurricane."

These wildly optimistic statements impressed the city's Chamber of Commerce, which gave Abbe the money to set up a three-month trial telegraphic weather reporting service. Perhaps they, like Abbe, were impressed with the track record for science set by astronomy. On September 1, 1869, Abbe began his Daily Weather Bulletin. It wasn't an auspicious beginning—at first only the St. Louis and Leavenworth, Kansas, observers wired in their observations. Nonetheless, the next morning the *Cincinnati Commercial* newspaper printed Abbe's report. Projecting the weather eastward, Abbe correctly figured that high pressure in St. Louis meant clear conditions over Cincinnati the next day, with a storm probable to the south. "Clouds and warm weather this evening. Tomorrow clear," was his confident report. But privately, he wrote in his diary, "First prognostication. A Doubt and a Failure."

Eventually 15 stations reported regularly, and Abbe scaled back his hopes of three-day forecasts to two days. Forecasts didn't appear every day, but the Chamber of Commerce was satisfied nonetheless. Sponsor Western Union extended the trial weather bulletins, but the scheduled opening of the Signal Service network doomed Abbe's own project. Upon joining the Signal Service project as its chief scientist, Abbe became a fixture in Washington. His success as a forecaster and crusader for meteorology over the next three decades earned him the nickname "Old Probs." The Grant administration appreciated the economic and publicity value of Abbe's high-profile services. By 1872, he was the second-highest-paid civil servant in the federal government, after Joseph Henry.

In contrast to James Espy, who was Washington's resident meteorologist from 1842 to 1857, Abbe was a cautious and serious presence in the nation's capital. He taught at Columbian College (now George Washington University) and founded a library and laboratory at the Signal Service headquarters. He also encouraged research within the Signal Service ranks and later among the Weather Bureau staff when the forecasting office moved to the Agriculture Department. Abbe also helped speed the progress of American meteorology by translating important European weather studies into English. He even was a cofounder of the prestigious Cosmos Club, which has long counted the capital's most distinguished thinkers as its members. Respected and well remunerated, Abbe fortunes were a stark contrast to those of the first forecasters of Europe.

Le Verrier and FitzRoy met considerable opposition in the 1860s. Le Verrier's telegraphic network wasn't as big a success as he thought it would be. The Academy of Sciences began to attack his methods in the 1860s, and in retrospect there was good reason behind their caution, even if politics more likely motivated the criticism. The climatologist Helmut Landsberg, who investigated the quality of those early forecasts a century later, wrote that the task Le Verrier and FitzRoy undertook "would leave a modern meteorologist groaning. Unstandardized observations, unreduced barometer readings, no upper air data, and large gaps in the observational network make us realize that our predecessors were really enthusiasts and bold prophets." For all the promise engendered by the telegraph and the new insights into storms, meteorology at midcentury was seriously flawed. Storms were still dangerous, and yes, they were still unpredictable.

FitzRoy paid the dearest price of all for his work as a forecaster. He was devoted to science throughout his distinguished career, but this devotion sometimes clashed with an equally strong sense of moral duty. In 1831, as captain of the *Beagle*, charged with surveying the world, he thought it worthy to take along an astute naturalist, his friend Charles Darwin. Like FitzRoy, Darwin maintained a keen interest in weather—he corresponded with Redfield about waterspouts. FitzRoy would later distance himself from Darwin, unable to reconcile evolution theory with his fundamentalist faith. In London, FitzRoy found himself caught between the cautious restraint of the British meteorological community and the obvious desire of the public for storm forecasts. While the charter for the Met Office called only for observations, FitzRoy felt compelled to try to save lives with storm forecasts. The scientific challenge tempted him as well. "It should always be remembered," he wrote, "that the state of the air foretells coming weather, rather than indicates weather that is present." With a telegraphic service in place, and observations finally rolling in, the temptation to forecast increased. Henry's example at the Smithsonian also must have encouraged FitzRoy, though the American had an important advantage. Since midlatitude storms basically move west to east, observations from the Plains told Henry what weather was approaching. But FitzRoy faced nothing but open seas and empty charts to his west.

In 1859, Britons mourned the loss of the steamer *Royal Charter*, which sank in a storm, drowning nearly its entire crew. FitzRoy received permission to warn coastal residents of storms in progress. He took the opportunity and ran with it. In February 1861 Fitzroy sent his first storm warning over the wires; six months later he was transmitting forecasts every day. FitzRoy knew the risk he was taking. Public need and interest were great, but meteorology was hardly up to the task of forecasting. For starters, most observers sent him data only once a day (at 8 a.m.); only a handful transmitted data again at 2 p.m., unless conditions obviously warranted extra reports. With this smattering of information, and with a blind spot to his west, FitzRoy tried to forecast two days ahead. A more careful man, like the preeminent French meteorologist of the day, François Arago, would have refused to do it. "Whatever may be the progress of the sciences," Arago was convinced, "never will observers who are trustworthy and careful of their reputations venture to forecast the state of the weather."

The English are a seafaring nation; at the time they lived and died by the sea. Many of them appreciated the warnings and FitzRoy's display of seaman's guts. Officials were less amused. FitzRoy carefully termed his forecasts "opinions," but scathing reviews of them appeared in *The Times*. His critics considered FitzRoy's forecasts voodoo, not science. Burned by the criticism and abandoned by his professional peers, FitzRoy slit his throat on April 30, 1865.

One of the sad ironies of the suicide of this first great forecaster of storms was that, in fact, his science was more than impeccable. Like his commitment to forecasting, FitzRoy's theoretical ideas were far ahead of his times. His method is outlined in his popular *Weather Book*, published in 1863. The charts FitzRoy used to describe midlatitude storms strike the modern eye immediately. Warm and cool airstreams wrap into a central vortex, forming interlocking commas of contrasting flow that look very much like idealized sketches of the storm cloud structures we now see daily from satellite cameras in space. Certainly, FitzRoy was not the most advanced theoretical thinker of his day. He did not have enough observations, mathematics, or reasoning to fully develop a theory from these ideas. But he had the intuition many others lacked. He had seen half a century into the future of storm science, but was martyred for his vision.

CHAPTER FOUR

An Enlightened Confusion

An ominous shadow swept over New Brunswick, New Jersey, at half past 5 on a sultry spring evening. "A very dense and low cloud stretched itself along for some distance like a dark curtain," recalled Professor Lewis Beck, who was riding a steamer up the Raritan River a few miles away that day in June 1835. A fellow passenger grabbed Beck and pointed toward a "black and terrible column" emerging from the sky. The funnel stretched toward the ground, merging with a cloud of dust swirling up to meet it. "In a few minutes the well defined character of these united cones was changed, and there arose a column, spreading at the top, and resembling a volcanic eruption. A vast body of smoke, as it seemed, rose up and again descended...."

The tornado was a relentless whirl of wind 300 or more yards wide spinning branches, dust, and roofs high into the air. As the storm mauled New Brunswick, helter-skelter explosions and a wafting smell of sulfur confused the frightened residents. Local firemen frantically rushed to and fro looking for the source of the smoke and the mysterious cataclysms, but they found no fire.

After plowing through New Brunswick and Perth Amboy, the tornado dissolved into a downpour of hail and heavy rain at the coast. In its wake, shattered buildings and scattered trees attested to the tornado's furious winds. Inevitably, in what has become an American tradition, leading scientists swarmed to the scene to test their theories about storms. Among the storm sleuths who visited New Brunswick were Alexander Dallas Bache, a descendant of Ben Franklin; Joseph Henry, recently settled into his post at Princeton; and Robert Hare, a prominent Philadelphia chemist.

The most important visitors were William Redfield and James Espy. The latter, the future first meteorologist of the federal government, would become the popularly anointed "Storm King" of his time. The two men were hot on the trail of vastly improved ideas about the circulation of storms. Redfield thought their work carried on the noble tradition of American meteorology. As he wrote to Espy a few years later:

> To establish the bare fact of the translation of storms in space and their continuity in time, you are doubtless aware, is little more than to complete a work that was commenced by Franklin nearly a century ago. Our personal interests or claim in these matters are, however, of small moment, and must await the awards of the future.

Noble sentiments, but the results of the New Brunswick surveys showed that storm science in America was no civil, selfless pursuit in the 19th century.

Redfield and Espy scrutinized every inch of the twister damage. They picked over the same fallen tree trunks, mapping the angle and order in which they fell. They saw the same piles of bricks, the same trails in the mud and manure, the same overturned outhouses. Yet the two men could not agree on what had happened, for Espy and Redfield fought one of the most heated, notorious, and ultimately personal disputes in the history of science. In New Brunswick they aimed to prove who was right about storms, one man at the expense of the other. Not surprisingly, each drew entirely different conclusions about which way the winds had blown. Espy believed that winds in all storms, including tornadoes, converged into the center like the spokes of a bicycle wheel. Redfield thought that tornadoes were small versions of hurricanes, generating intense winds that circle a calm core. Espy insisted that the New Brunswick damage fell or blew into the center of the tornado path, proving that Redfield was deluding himself. Alexander Bache found evidence for Espy's claims, which galled Redfield, who had spoken to eyewitnesses who swore the pillar of wind rotated. Redfield also had uncovered in his survey "numerous facts which appear to demonstrate the whirling character of this tornado."

Because Redfield and Espy's views seemed incompatible yet compelling, prominent scientists took sides in the debate. "Indeed, meteorology has ever been an apple of contention," Henry remarked, "as if the violent commotions of the atmosphere induced a sympathetic effect in the minds of those who have attempted to study them." The dispute revealed the immaturity of the science: meteorologists lacked a common education and common quality measurements. The observations were vague enough to support practically any theory. Indeed both Espy and Redfield were wrong, and both were right, at the same time. Both men contributed essential concepts to storm science. Meteorologists eventually sorted out

By drawing stylish vertical striations in these waterspouts, the illustrator for an 1869 meteorological text followed James Espy's theories, contradicting the now well-observed spiral flow in the vortex. NOAA photo.

the right from the wrong in the Espy–Redfield dispute, paving the way to the tremendous breakthroughs awaiting in the next century.

Amid the confusion of the Espy–Redfield dispute, storm science found new enlightenment in the 19th century. Meteorology emerged with central concepts based on laws of heat, thermodynamics, and gases. This new core of knowledge spun off the field's first quantifiable theories. Physical concepts became the language of meteorology. Today, not only theorists but also storm chasers, rainmakers, and many TV weathercasters speak it fluently. Espy, much more than Redfield, fostered this breakthrough. Meteorology was ripe for an infusion of valid theory, and Espy provided it. His ideas set the tone for the future of storm science.

James Espy was a native of Pennsylvania but spent his boyhood in Kentucky, eventually graduating from Transylvania University in Lexington. He became a schoolteacher, and in 1820 joined the faculty of the Franklin Institute in Philadelphia. This was a seat of power in early American science, but originally Espy was no scientist—he was a language instructor. At the institute, however, he began to study meteorology and the behavior of gases. He found the experiments of John Dalton particularly inspiring. Dalton was the Englishman who formulated the first modern theory of atoms in 1805. A close friend of cloud guru Luke Howard, Dalton showed a relationship between expanding, cooling air and the condensation of the water vapor within it. This lead Espy to a new theory of rain, and then of storms as a whole.

With Espy's storm theory, meteorology finally made good on the promise of the barometer, invented nearly 200 years earlier. The physics of air advanced quickly after the development of this crucial instrument. The barometer gave the atmosphere a new, visceral meaning. Torricelli and others realized that the air in the atmospheric "column" overhead creates pressure by virtue of its weight. Every moment we spend on this planet (near sea level), more than 14 pounds of force presses constantly against every square inch of our bodies. Otto von Guericke vividly demonstrated that omnipresent weight of air in the mid-17th century. He pumped out the air between two matching metal hemispheres, thus sealing a vacuum inside by virtue of the crushing pressure of atmosphere surrounding the metal. Then he had two teams of four horses play tug-of-war with the hemispheres in an attempt to pull them apart. The horses snorted and strained, but to no avail. The air pressing the hemispheres together was too strong.

The discovery of atmospheric weight immediately suggested a corollary: atmospheric pressure should decrease as you climb higher, because the column overhead is shorter. In 1648 the French mathematician Blaise Pascal set up a barometer and thermometer and then sent his brother-in-law, Perier, with similar instruments in hand, to climb by foot to the highest point in the district, Puy-de-Dôme. The two men took readings at a prearranged time and found, as expected, that the pressure at the bottom of the mountain was greater than at the top. Perier must have been in good shape: he was so intrigued by the result that he switched barometers and hiked up the 4,800-foot peak again to confirm the results. Nor was he through. The next day Perier recorded a tiny drop in pressure by scaling the highest church steeple in town.

Discovering the reduction—or "lapse"—of pressure with increasing altitude was only the first step toward Espy's theory. Next, scientists began to find that gases behave according to laws. In 1660, English philosophers Henry Power and Richard Towneley showed that, if you can expand a gas without changing temperature, the pressure decreases. French physicists Jacques Charles and Joseph Louis Gay-Lussac showed that temperature also decreases as a gas expands. In 1801–2, with detailed measurements, Gay-Lussac was able to get very specific about this law: expand a gas—any gas—by 3.75% and it cools by 1 degree Celsius. Therefore, let (dry) air rise and lose pressure, and it will cool at a quantifiable rate.

By the middle of the 18th century, natural philosophers began to use these concepts. They suggested that surface heating spurred lighter air to rise, causing upper-level air to redistribute, thus lowering pressure over the heated area. Franklin later mused that waterspouts—the tornadolike whirls over water—form around a core of heated rising air. In 1735 the English barrister George Hadley proposed the classic weather theory invoking rising air. He invoked three basic meteorological concepts—convection, rotation, and surface friction—to explain why sailors could reliably find easterly winds in the tropics and persistent westerlies at midlatitudes. First, Hadley described the motion of the atmosphere as an oversize cell of convection that works the same way as overturning water in a pot of boiling water on the stove. He said that heat in the tropics rarifies the air, which then rises, forcing colder, denser air to flow near the surface toward the Equator. The tropical air, meanwhile, spreads poleward at high altitudes before sinking back to the ground.

Second, Hadley said the Earth's eastward rotation twists this heat-driven circulation into easterlies and westerlies. At the Equator, still air actually travels about 24,000 miles a day just to keep up with the rotating Earth. That's 1,000 m.p.h. eastward. But the rotational speed of the Earth's surface (with its air) decreases toward higher latitudes; it is basically zero miles per hour at the poles. Hadley claimed that the slow air moving Equator-ward turns westward relative to the Earth as it "falls behind" the faster-rotating surface. These are the easterly trade winds that propelled Columbus and Magellan during their voyages. Conversely, as 1,000-m.p.h. tropical air moves poleward, it appears to curve toward the east to form midlatitude westerlies. Hadley calculated that air moving from the Tropic of Cancer to the Equator would acquire a relative easterly movement of 80 m.p.h. Yet he argued that this was impossible, because he believed that no wind, not even a storm, could exceed 60 m.p.h. He explained the discrepancy with a third key concept of meteorology—the friction holding back surface winds. Every bump from mountains to ant hills drags the air, keeping it from slipping to superhigh speeds.

Hadley's three points about the general circulation of winds eventually spread to storm science, too. Convection, in particular, caught many scientists' attention. In 1816 a member of the Berlin Academy of Sciences declared, "The principle of the ascending current of air should really be called the key to the whole science of meteorology." Espy made this a reality. He matched convection with a crucial concept discovered in 1757 by the Scottish scientist Joseph Black, a consultant to the inventor of the modern steam engine, James Watt. Black noted that a fire quickly heats water to its boiling point, but afterward, despite more and more heating, the temperature holds steady as water slowly turns to steam. The fire, Black argued, must somehow transfer its heat to the water in a form unaccounted for by temperature. In other words, water in gas form has more heat than water in liquid form, even at the same temperature. The difference is latent heat, which water must absorb to become steam. The reverse must be true as well, Black argued: when vaporous water becomes liquid water, it releases a tremendous amount of heat. The conversion from ice to water and back holds similar wonders, said Black. Relentless spring sunshine does not melt snow instantaneously at 32 degrees Fahrenheit, causing disastrous floods, because much latent heat must be added to the ice first.

Espy brought latent heat to its rightful home when he used it to form his meteorological theory. Storms are almost all violent byproducts of the daily, global recycling of water from atmosphere (gas), to oceans (liquid) and glaciers (solid), and back to atmosphere. It was through this water cycle that Espy entered the contentious arena of storm theory armed with the knowledge of latent heat. He used the concept to attack the prevailing theory about how rain and clouds form. At the time, some scientists believed that the mixing of cold and warm air caused clouds and rain. Scottish geologist James Hutton suggested this in 1784 based on common experience: on a very cold day, warm breath condenses into a cloud when exhaled. More vapor can exist in the warm air inside the body than in the wintry atmosphere outside, and breath obviously cools when it mixes with the wintry atmosphere. So Hutton figured that combining cold air and warm air in proper portions yields a mixture of air in which some of the moisture originally evaporated in the warmer air must condense, forming a cloud and eventually rain.

Espy set the record straight by experimenting with something he called a "nephelescope"—a glass vessel in which air pressure could be varied with a pump. In 1828 he used it for his first scientific paper, on heat and the expansion of air. Then he began investigating moisture. Espy weighed a sample of air saturated with water vapor, then allowed it to expand. This cooled the air, whereupon the moisture condensed out. Espy then pumped the now-dry air back to its original pressure and weighed it again. Amazingly, the drier air was heavier than the saturated air. "The result was an instantaneous transition from darkness to light," Espy later recalled. He could see that moist air might rise by virtue of having lighter water vapor in it.

The nephelescope also enabled him to measure the change in heat in the air as vapor becomes cloud. He knew from Gay-Lussac that as air rises, it expands and cools. Moist and dry air cool at the same rate if they rise together—but only at first. The nephelescope showed that when vapor in the moist air begins to condense, the expanding moist air no longer cools as fast as the expanding dry air. Latent heat release in the moist air mitigates against much of the expansional cooling. Thus, latent heat release can make rising moist air warmer and lighter than its surroundings, so the moist air continues to rise and cool, forcing more and more vapor to condense. Espy correctly concluded that rising moist air—not cold air mixing with warm air—accounts for the condensation that makes most clouds and rain.

Espy was borrowing a relatively new idea from Dalton and Gay-Lussac. The Europeans had investigated a special case of gas activity—that in which no heat is exchanged between the gas and its surroundings. This "adiabatic" case is easy enough to produce in the lab, but Espy made an important leap by saying that such conditions closely approximate the way air expands and contracts in nature. It is always a treat for scientists when nature follows the ideal case, and that is what happens when air rises and falls. Adiabatic expansion and contraction were, Espy enthused, the "lever with which the meteorologist was to move the world." While Espy experimented with the nephelescope, Pierre Simon de Laplace and his student, Simeon Denis Poisson, formulated the adiabatic expansion mathematically. Meteorologists later rediscovered their adiabatic equations, which remain at the foundation of storm study today.

Espy's theory gave storm watching new meaning. It showed that the height of a cloud's base often indicates the humidity of the air. When air begins rising, the moisture in it is vapor. It is invisible. But the altitude at which droplets begin to appear—the bottom of the cloud—is where the air has expanded enough to reach a low enough temperature for saturation to occur. The temperature at which saturation will occur (called the "dewpoint") is lower if the air is drier. Lower the dewpoint of a pocket of air 1 degree, and the air probably has to rise 300 feet higher to condense. Thus high cloud bases (cool saturation temperatures) mean dry air, and low cloud bases (warm saturation temperatures) mean moist air.

Two years after Redfield's remarkable first paper on the circular nature of winds in storms, Espy charged into storm theory full bore, extending his rain theory to propose that storms form when air somewhere is lighter than its surroundings. This could happen when air is laden with water vapor. Or it could occur because the air expands from heat. In either case, the air rises. If it reaches an altitude where it cools to the dewpoint, clouds form. The release of latent heat makes the air even lighter, driving it upward even faster. The chimney of air eventually spreads out at high altitudes before sinking outside the cloudy storm core. This evacuation of air from the chimney lowers surface pressure below. Espy said winds at the surface then blow directly into the center, feeding the updraft, making the storm a giant convection cell.

Redfield also was intimately familiar with latent heat and the expansion of hot gases. Steam power prompted many scientific investigations into these matters, and Espy himself compared his storm model to a steam engine. Redfield's Hudson River shipping business in New York, meanwhile, ran on steam power. He even won a name for himself in the business by advocating "safety barges"—a system in which a steamboat towed passengers on a barge at a safe distance behind the volatile engines. Redfield was enough of a believer in steam engines to urge the establishment of a railway across the Appalachians (on a route he reconnoitered on his walk to Ohio in 1810). Somehow this awareness of steam power didn't make Redfield a believer in Espy's theory. Instead, he was its most bitter opponent.

In 1834 the inevitable clash of storm ideas erupted. Espy came across Redfield's whirlwind theory and saw intolerable discrepancies. From the start, he felt Redfield was unscientific. Some of Redfield's ideas, Espy wrote in 1835, "are so anomalous and inconsistent with received theories; that I hesitate to put entire confidence in them, and shall continue to doubt until I have the most certain evidence of the facts." In one of his most damaging claims, Espy wrote that Redfield had a poor understanding of atmospheric tides over the Pacific. If Redfield had any explanation at all for storms, it was based on regular tidal activity in the atmosphere rather than on heat or other forces.

Redfield seized on these comments in a published response deriding Espy for following Hadley's old ideas: "The grand error into which the whole school of meteorologists appear to have fallen consists in ascribing to heat and rarefaction the origin and support of the great atmospheric currents which are found to prevail over a great portion of the globe." The idea that a central updraft would cause wind to rush in from all directions was also preposterous, Redfield claimed. "Concentrating wind from every point in the compass" would collapse the system.

In a generous mood, Redfield believed that Espy ignored vortex patterns because he studied the wrong storms. Snowstorms, for instance, don't generate a classic spiral anyway, Redfield conceded. Espy responded by modifying his theory: winds could converge toward a horizontal axis within the storm, rather than a single central point. Nonetheless, heat would be the driving force. Then Espy attacked Redfield's 1821 survey and claimed that centrifugal forces would destroy the tight circulation Redfield proposed.

Into the fray entered Robert Hare, insisting that both Espy and Redfield were wrong. He believed electricity in the air caused whirlwinds. This idea descended in part from Benjamin Franklin's work on atmospheric electricity, but whereas Hare was a brilliant chemist, as a meteorologist he was just another thorny personality in an already prickly debate. In 1844, aware that Hare was on the scientific warpath, Redfield wrote resignedly to the eminent German meteorologist Heinrich Dove that Hare was a "pertinacious antagonist" and would rather win his argument than shed "any new light on the subject of controversy." Redfield chaired the first meeting of the American Association for the Advancement of Science in Philadelphia in 1848 and took advantage of the occasion to introduce Henry Piddington's *Horn Book of Storms* to his peers. Hare "immediately rose" to deny the possibility of rotation in storms. And in yet another AAAS meeting, Hare attacked Espy and Redfield for ignoring the "all powerful agent," electricity, and challenged Espy to a debate before Congress.

While many found Hare eminently disagreeable, Espy and Redfield at least made a few attempts to meet cordially. Redfield stopped by his rival's home in Philadelphia once, only to find that Espy was away. Espy invited Redfield to Philadelphia to observe activities of the Franklin Kite Club. (Kites were no more an idle hobby for Espy than for Franklin: Espy used kiteborne thermometers to verify his cloud height calculations. He was pleased to see that the kites shot skyward in the updrafts, as he expected.) After a small tornado in Newark in 1839, Espy invited Redfield to venture across the Hudson so the two could investigate the storm track together.

Redfield declined, saying he was too busy, but added, "I shall rejoice at any actual advance you may accomplish in meteorological science." Espy undoubtedly bristled at the sarcasm.

Redfield was in fact a scientist only in his spare time. He also never lectured publicly on his storm theory—not even at scientific meetings. Denison Olmsted, his Yale mentor, took up the cudgel for him on occasion, meeting Espy for a debate on storm theory in Boston in 1840. Espy, on the other hand, was an ardent popularizer, and his notoriety was critical for the later success of meteorology and of forecasting in America. Journalists, at least, were impressed by his presence. One newspaper reported that during his debate with Olmsted, Espy "stood like a rock in a midst of his storms; and though his manner was firm and undaunted, it was dignified and respectful to his opponents." Another swooned, "The name ESPY may hereafter stand ... high upon the list with Galileo, Harvey, and Franklin...." Espy took his theory—and his nephelescope—on the road, lecturing on the Lyceum circuit beginning in 1837. He also published a series of pamphlets on meteorology and rainmaking in the late 1830s. Meanwhile, Redfield stewed mostly in private, clearly obsessed with Espy's growing fame. He kept a file of newspaper articles about his nemesis's lectures.

Espy may have reached out to the public in part because he was a professional: he had to find a way to make money in meteorology. His popularizing efforts helped win him the nickname "Storm King." Other scientists were not amused by the coronation, partly because Espy was lecturing on the possibility of making rain artificially. Most meteorologists considered the subject akin to alchemy. Espy's ideas on the subject were as grandiose as his debating demeanor. He suggested torching a line of forestland hundreds of miles long to produce a concentrated zone of rising air. With the attendant latent heat release due to cloud formation, Espy could imagine a massive confirmation of his storm theory. In 1849 he set fire to 12 acres of pine in Fairfax County, Virginia, to test his ideas. The blaze was a futile demonstration of self-aggrandizement in the name of storm science.

In 1842, the War Department appointed Espy Professor of Meteorology. At first the Army asked him to review meteorological records collected by the Surgeon General, but soon he was working for the Secretary of the Navy, issuing reports on meteorology to Congress. In the early 1840s, after collecting his papers into the influential book *Philosophy of Storms*, Espy was able to experiment with making thrice-daily synoptic maps based on military data. Meanwhile, Redfield, who never published a book of his own, tried to get the U.S. Navy to follow the Royal Navy's example and buy for its officers copies of William Reid's work on rotary storms. With Espy in a seat of power, that never happened. Instead, in 1851, the U.S. government published the Storm King's own rules on evading storms.

When Joseph Henry arrived in Washington to take over the Smithsonian, Espy's monopoly on government meteorology ended. The Army transferred Espy to the Smithsonian to work under the brilliant, diplomatic Princeton professor. Espy mapped results from the new observation network. The assignment was appropriate: Henry started the Smithsonian network in large part to settle matters between Espy and Redfield. He wisely believed that someday their ideas, "contradictory as they may now appear," would end up both contributing to storm knowledge.

By the 1850s, Espy had backed away from theory and returned to experiments, having refined his nephelescope. He retired and went to Ohio in 1857, spending the last three years of his life there preparing manuscripts on "moral accountability." To his credit, Espy was an ardent voice for government involvement in storm warnings. Unfortunately, congressmen who heard his pleas also could not help but notice his ego. One called Espy a "monomaniac" whose "organ of self-esteem was swollen to the size of a goiter." His friend, Alexander Bache, was more charitable: "His views were positive and his conclusions absolute, and so was the expression of them."

Redfield and Espy each had glaring faults. Redfield's theory lacked any significant connection to precipitation. And Espy's theory yielded to a typical temptation, proclaiming one basic cause for all clouds and rains. Strip away the intransigence and generalizations, however, and an essential philosophical polarity emerges between Redfield and Espy. In general, one could say Redfield was right that violent storms such as hurricanes and tornadoes basically have spiral winds. And Espy was right that, in essence, storms are rising air fueled by the condensation of water. In other words, Redfield was the superior observationalist, and Espy was the superior theorist.

Ironically, this difference in scientific method fueled their debate. Redfield and Espy disagreed on what science should be. Redfield generally refused to explain rotation in storms. When he did so, he flailed about hopelessly. He believed that the goal of science was to reveal natural laws, the correlations and patterns in real-world data. To explain such laws was pure speculation, too subjective for a scientist. Espy, on the other hand, tried to explain natural law. He believed scientists would not advance without using hypotheses to guide their experiments and observations. Redfield abhorred Espy's findings in part because Espy first made a hypothesis (based on experiments) and then went out and collected data from real storms to support it. The Espy–Redfield debate was thus part of a schism in ideology that affected many scientists of the day. Among American meteorologists, the accuracy of Redfield's observations stymied theoretical advances for decades.

As synoptic charts improved enough to reveal storm winds clearly, Espy's radial model lost credibility. Unfortunately many scientists forgot his valid points about latent heat and adiabatic change. Elias Loomis, however, was an exception. A pastor's son, Loomis (1811–1889) was one of the best-trained meteorologists of the era. He studied math and physics at Yale, under Redfield's mentor Olmsted. Then he went to Paris to study with the master, Arago. Eventually, after posts at Ohio Western Reserve College and the University of the City of New York, Loomis made it back to New Haven as professor of natural philosophy and astronomy. The Storm King was lucky to have such a worthy prince supporting his ideas.

Ideologically, Loomis took after Redfield more than Espy. Even when he made a discovery, Loomis was loath to explain it. In short, he was everything that Espy was not as a scientist—a patient observer first, a cautious theorist last. He was devoted to collecting data and did it with patience and thoroughness. For one project relating magnetism and temperature, he read his thermometer once an hour from 6 a.m. to 10 p.m. every day. He collected enough information to map the declination of the compass needle over 13 states. The map required calculating and smoothing averages by hand for more than 4,000 readings. Storm science needed a dogged researcher like this to do the impossible: reconcile the contradictions Espy and Redfield hurled at each other. Fortunately, Loomis was game to try.

The work of Elias Loomis represents the synoptic method in one of its finest hours. Loomis began using maps to show several meteorological factors at a glance and opened the door to modern extratropical storm theory. Like Brandes, he drew lines of equal deviation of pressure from normal, as well as indications of cloud cover, wind direction, and other factors all on one map. (Isobars—lines of equal pressure—were adopted later by forecast services.) Loomis believed synoptic maps should show "nearly every circumstance essential to a correct understanding of the phenomena of a storm ... to the eye at a single glance."

Fortunately for Loomis, data meeting his strict requirements were available for some storms. Weather information had been gathered during the solstices, beginning in 1835, because some scientists organized an international effort to investigate the effect of celestial events on the atmosphere. During one of those solstices—December 20–23, 1836—a strong winter storm swept across the Midwest and into the northeastern United States. Loomis's careful study of the data showed that both Espy and Redfield were right: winds indeed rotated counterclockwise around the low, but they also converged inward at the same time. Taking a page from Hadley, Loomis explained that Earth's rotation had deflected the winds. The difference between Espy and Redfield was largely this rotational influence. Now Loomis could reconcile Espy's theory to the synoptic data nearly as easily as Redfield's.

Loomis actually went much further in his storm studies. Both Espy and Redfield depicted symmetrical storms. But Loomis's more careful synoptic analysis showed a startling east–west contrast south of the low-pressure center. Cold, northwesterly air dominated west of the low center, while warm, southerly air dominated east of the low. For Loomis, the clash of warm and cold air currents clinched the importance of rising air, which was, he said, "at least in this latitude, the most common cause of rain." He explained, "When a hot and cold current, moving in opposite directions, meet, the colder, having the greater specific gravity, will displace the warmer, which is thus cooled and a part of its vapor precipitated." In Hutton's day, Loomis would have been tempted to say the cold and warm air mixed to form the rain. Thanks to Espy, he instead realized that cold dense air scooped under and lifted the warm air.

Unfortunately, the time was not ripe for attributing storms to the clash of cold and warm air. Espy's explanation, relying on the release of latent heat through convection, was at once experimentally sound and seductively simple. No meeting of airstreams was necessary. Understandably, Loomis fell under its sway, abandoning temperature contrasts to future generations of meteorologists. By 1846 latent heat was the centerpiece of his storm theory, and Loomis was Espy's most influential advocate. Loomis included temperature contrasts in his theory of the storm life cycle, but not as the instigator. First, he stated, an area of unevenly hot or moist air disturbs the generally westerly flow. Then, clouds form and latent heat release strengthens updrafts. The resulting convergence causes a cool, northwest wind to cut under the warm southerly wind. The converging winds from all directions begin to rotate around the center, spiraling inward. The storm is embedded in a westerly flow, so faster westerly surface flow presses against the southeasterly currents. This compression eventually wipes out the low-pressure area.

Loomis moved Americans temporarily ahead of European storm theorists, many of whom were practically unaware of Espy while following the ideas of the great Prussian meteorologist Heinrich Dove. Dove's methods were dated from the start. Brandes, Dove's teacher, had established the synoptic method, analyzing a broad area of stations for a single moment. Dove rebelled by studying changes at one station over time. Dove tracked the wind shifts and pressure changes at Konigsberg in September 1826, and found "a remarkable phenomenon." Pressure began falling as the storm first approached, and winds began veering around the compass. Even as easterlies predominated at the surface, cirrus rode southerlies aloft and eventually these southerlies swept across the surface. The cloud cover got thicker and thicker, and flakes of snow yielded to rain in the warmer conditions. Next the wind veered to southwesterly, the temperature rose, rain fell, and pressure bottomed out. Later, with the westerly, snow fell as pressure rose, followed by northerlies that brought clear skies.

The pattern, Dove believed, was evident in most storms, if not always as ideally as in the Konigsberg case. Dove came to understand the progression of winds and pressures as a tussle between warm currents from the Equator and cold air from the poles. The storm was merely the midlatitude meeting of contrasting winds, he said. Dove's ideas ruled European meteorology for the three decades he headed the Prussian Meteorological Institute in Berlin. When his power finally waned in the 1860s, meteorologists dropped Dove's ideas nearly completely. After all, Dove got the wind cart before the pressure horse: "The dependence of the barometric pressure on the wind prevailing at the time of observation confirmed without exception that it seems suitable to regard barometric pressure as a function of the wind direction." The truths in Dove's theory would have to wait a few more decades for a revival.

In America, by contrast, storm theory continued to move ahead because the next great meteorologist added new mathematical rigor to the Storm King's ideas. Like Redfield, William Ferrel was a self-taught man. Like Espy, Ferrel began his career as a schoolteacher—in Nashville, Tennessee, in the 1850s. He became interested in the science of weather when an editor asked him to review a book by Matthew Maury on world wind patterns. Ferrel saw so many theoretical flaws in Maury's book that he felt the only proper response was to do his own research. Fortunately, his interest in mathematics was strong. Soon Ferrel had developed the first complete mathematical theory of how the Earth's rotation affects the motion of air over it.

Unlike Hadley, who considered the effects of rotation only on north and south winds, Ferrel could prove the effects on all winds, including east and west currents. Laplace in 1803 and Gustave Gaspard Coriolis in 1835 had already established this mathematically in other fields, but Ferrel set forth equations for this effect as it applied to weather. His first paper in the field, "An essay on the winds and the currents of the ocean," appeared in the rather obscure *Nashville Journal of Medicine and Surgery* and not many meteorologists saw it. Certainly, few storm theorists would have seen what was coming from Ferrel. Ferrel believed Espy and Loomis were right that the center of the storm was an updraft of light, warm air driven by latent heat release. And like Loomis, he believed that the converging winds around the low-pressure core of the storm were deflected into a cyclonic whirl by the rotation of the Earth. The low center was surrounded by a belt of high pressure caused by sinking air, just as Espy claimed, cutting the storm circulation off from the surrounding atmosphere.

Ferrel was also the first to study the upper circulation of storms closely. He pointed out the consequences of the heated column of air at the center: the storm simultaneously has a low central pressure at the surface and a high central pressure at the top. A hotter, rarefied column expands, becoming taller than an adjacent cooler column. Somewhere, then, there is an altitude at which the warmer column has much more air above it—meaning greater pressure—than the cooler, shorter column. The upper high pressure explains the upper outflow of the storm. Since outflow is deflected by rotation, a northern hemisphere storm, which rotates counterclockwise at the surface, will simultaneously rotate clockwise at high altitudes.

Ferrel's symmetrical model of a storm basically holds true only for hurricanes where there is spiral inflow at the bottom and spiral outflow at the top. Nonetheless, his work, like Espy's, represented a growing confidence in the importance of a few key properties of the lower atmosphere. Gravity makes most air molecules sag toward the ground: half the air molecules lie below 18,000 feet. About three-quarters of the mass of the atmosphere lies within the bottom 30,000 to 60,000 feet, the weather-making layer we now call the troposphere. The troposphere, our sphere, makes storms because it is heated by the surface of the Earth below and because it circulates the vast majority of water found in the atmosphere. Higher layers in the atmosphere discovered after Ferrel's time—the stratosphere, the ionosphere, the mesosphere—all lack one or more of these qualities. Water and heat from below are thus the keys to storms. These concepts were the path Espy established. They turned meteorologists' attention toward the vertical, a dimension largely lacking in the earlier scientific theories of storms. Theorists became more confident that the vertical dimension, not the mundane horizontal winds, held the key to severe weather. Not surprisingly, Espy's ideas flourished and expanded in a new aeronautic age that had as remarkable an impact on our view of the world of storms as did our first glimpses from space satellites in the 1960s.

Today, we look up, see towering cumulonimbus to 60,000 feet, and have some idea what that means, because we can fly to more than half that height in a jet airliner. We know the strongest storms are taller than Mt. Everest. But for many centuries, right up to Espy's time, altitude was a nearly meaningless dimension. In the Classical world, Olympus was the king of the mountains at less than a quarter the height of Everest, and at its top the weather was believed to be perfectly calm. As late as 1822 storm theorist Robert Hare published his thoughts on coastal storms and claimed that the clouds of the Gulf of Mexico could never rise above 5,000 feet. Such ignorance of the depth of weather hampered the development of ideas about storms.

But inevitably, new technology turned the tide of meteorology upward, beginning in France. Delighted with the way heated air made sacks float, two paper manufacturers in Annonay took a logical but revolutionary step. On a rainy Wednesday—June 4, 1783—Joseph and Etienne Montgolfier unveiled a 28,000-cubic-foot paper bag, its giant sheets held together by 1,800 buttons. A blue-ribbon panel of local functionaries watched as a fire hauled by the contraption heated the air inside enough to lift it 3,000 feet into the air. The first balloon rode the winds for a mile in a leisurely 10 minutes before burning up on landing. From the ashes of the Montgolfier's invention rose a whole new way to study the stormy moods of the atmosphere.

As they would later do with everything from the telegraph to the computer, meteorologists were quick to take advantage of the new technology. Dr. John Jeffries was a Bostonian, Harvard class of 1763. But he was also a Loyalist, and when the Revolution broke out, he was practicing medicine in England. Fortunately, not only for his Tory hide, but also for science, he stayed there. Not long after the war, the French balloonist Jean-Pierre François Blanchard was demonstrating the new wonder of aeronautics in London. Jeffries saw an opportunity to test ideas about the atmosphere and secured money from the Royal Society for an expedition. On November 30, 1784, he made the ascent with Blanchard, traveling from London to Kent in about an hour and a half. But destination was no matter: Jeffries was busy rowing a pair of oars in flight (not much help) and checking a thermometer, barometer, hygrometer (to measure humidity), and electrometer. He also brought stoppered bottles of water to empty and recork, thus securing samples of high-altitude air with the same collector's instinct that inspired astronauts to bring rocks back from the moon.

Every few minutes he checked the pressure and temperature and jotted the data down in the trip's log with a silver pen (rather than trust ordinary lead to the vagaries of the unknown sky!). For 100 years after, Jeffries' data were accurate enough to prove that temperature fell about 1 degree for every 360 feet in altitude up to 10,000 feet. He also found that moisture on this trip hovered near the bottom of the atmosphere, and that electric charge on this fair day did not vary with altitude.

Jeffries was bitten by the ballooning bug. The following January he and Blanchard took off from Dover, knowing that if the clouds tended to move east, winds at that height could make them the first to fly the English Channel. It is said Blanchard tolerated sharing the honors with Jeffries in part because the doctor put up the money. Blanchard made his human ballast promise to jump into the cold Channel waters if the balloon lost too much altitude. Not much new data came from the trip, because Jeffries, rather than fulfill his promises, wisely dumped his coat and scientific instruments into the drink when the balloon plunged precariously. He and Blanchard got a hero's welcome in Paris.

They deserved it. Not all scientists were brave enough to explore the atmosphere first-hand. Earth-bound scientists were smart, not cowardly. The first American airship, a 43-balloon craft built by Philadelphia scientists in 1783, took off with James Wilcox, a carpenter who had been hired for the task. Wilcox almost died in a rough landing. He was lucky. Ninety years after the first flight, the balloonist/meteorologist Flammarion counted 15 deaths in 3,500 scientific ascents. Like storm chasers today, early aeronauts faced obvious hazards. Unlike most storm chasers, they were ill prepared, often risking their lives. (Blanchard was a notable exception, having invented the aeronaut's perfect fashion accessory: the parachute.) Jacques Charles invented the hydrogen balloon, but then had to plead with Louis XVI for permission to fly it. The king, mindful of the value of one of his top scientists, had originally intended to use a convict as a test pilot. He relented and let Charles and another pilot fly on December 1, 1783. The spectacle delighted a crowd of 600,000 people, among them Benjamin Franklin, who was in town on diplomatic duties. After landing, Charles excitedly decided to fly again at dusk without his scheduled partner. He forgot that the balloon was now too light, and it shot up to 10,500 feet before he could stop the ascent. His fingers numb, Charles nonetheless became the first person to see the sun set twice on the same day.

Giddy with the sights, meteorologists were now on a quest to top the clouds. In 1804 Gay-Lussac flew solo to 23,000 feet above Paris and reported that the clouds were climbing still higher. The new perspective on the weather was immediate. Said one early balloonist:

> I shall never forget this remarkable procession of clouds sailing along with great rapidity below the car of our balloon. They were like a quantity of flocks of wool drawn along by some invisible hand ... their whitish cumuli appeared to issue from the surface of the ocean. How could fear, or any such emotion, find place in our thoughts when such novel and marvelous scenes were before us?

The greatest high-wire act of them all was meteorologist James Glaisher. Glaisher was a sort of early altitude freak, a precursor of today's fighter pilots and rock climbers. He flew more than two dozen times on scientific flights from Greenwich Observatory in England. Glaisher's crowning feat was a flight on September 5, 1862, from Wolverhampton. In less than an hour he and Henry Coxwell had topped 25,000 feet. This caused little trouble, and Glaisher continued his observations while Coxwell, working the instruments, began to huff and puff noticeably. In another hour, the temperature was still a rather balmy minus five degrees, but Glaisher began having trouble reading his instruments. As the barometer dropped below 10 inches, near 29,000 feet, Glaisher dropped his arm to the table beside him and found he couldn't move. "I dimly saw Mr. Coxwell, and endeavoured to speak, but could not. In an instant intense darkness came over me ... but I was still conscious," he recalled later, unsure when he blacked out. Coxwell found his arms limp beside him from the cold. In a last act of desperation, he pulled on the valve with his teeth, releasing enough gas from the bag to begin an uncontrolled descent. The carriage, the balloon vent, and Coxwell were covered in hoar frost.

This 1873 evocation of the peaceful pleasures of a world above storms accompanied James Glaisher's translation of a meteorological treatise by fellow aeronaut Camille Flammarion. By this time, however, both scientists knew quite well that rain often falls from multiple decks of clouds, thousands of feet thick.

Their breath and the moisture they carried up with them had frozen on them in the intense cold and low pressure.

When he came to, Glaisher checked his instruments and calculated they had risen to an amazing 37,000 feet, 8,000 feet higher than Mount Everest, without oxygen. The relative warmth the aeronauts encountered at the extreme altitudes helped save their lives, but it misled scientists for a few decades. In 1898, thinking that an 80-degree heat wave at the surface meant balmy ballooning at high altitudes, another pair launched from the Crystal Palace in London and endured an amazing 120-degree temperature drop in the 35 minutes it took to reach 27,000 feet.

But Glaisher was also a thorough scientist and genuine lover of the clouds, having spent months in the fog of the Irish high-altitude astronomical observatory on Keeper Mountain, near Limerick. Familiar with the crystalline environment of clouds, this first president of the Royal Meteorological Society waxed eloquently about the scene from his balloon: "[E]ndless variety and grandeur, and fine dome-like clouds dazzled and charmed the eye with alternations and brilliant effects of light and shade." In his book, *Travels in the Air*, he compared himself to a mountain climber. He got used to signs of extreme conditions. He knew he was at about 17,000 feet when his lips turned blue, at 19,000 feet when his hands became dark blue, and at 22,000 feet when his heart was audible. Glaisher showed that the largest lapse rates on overcast days were below the clouds—maybe 1 degree every 300 feet up to about 5,000 feet.

Flights into clouds were essential. As one aerologist put it, "What should we know of the ocean if a few sailors only had navigated at a short distance from the coasts, without losing sight of port?" Balloonists not only revealed the varying temperature profiles possible in the atmosphere, but they also showed the terrific winds in the upper atmosphere. Long before airplane pilots suspected the presence of the jet stream, balloonists searched for a current, like the Gulf Stream in the Atlantic, that would reliably whisk them across continents. Searching for this aerological grail, a French scientist sent nearly 100 paper balloons into the air at Amiens with return postage guaranteed. Sixty were returned, nearly all from east of the launch site. Some reached average speeds of at least 100 m.p.h. Other balloonist-scientists revealed to their Earth-bound colleagues that rain clouds do not always act alone. Sometimes the aerologists would emerge above a low cloud deck in rain only to discover the precipitation came from another deck above.

More puzzling, they found that the supposed steady cooling (lapse rate) with altitude was not true. In England, John Welsh of the Kew Observatory made numerous ascents in 1852 with a special thermometer built for the calm of balloon flight. Welsh rigged bellows at the bulb so that he could pump air across the thermometer. This ensured that the mercury responded to purely environmental air rather than to the warmer air in the carriage. On many days, Welsh rose through air with normal lapse rates to a significant altitude, only to find that the temperature inexplicably stabilized or dropped very little for a couple thousand feet. We now know that such a cap of relative warmth can hold back storm formation temporarily, because rising air cools too fast to continue ascending.

Lord Kelvin, one of the founders of thermodynamics, got interested in meteorology and helped analyze Welsh's puzzling finds. The science is

Limited to surface observations for routine forecasts, meteorologists of the 119th century had little evidence of such normal, yet bizarre wind shears with height as these seaside breezes sending smoke awry. Photo courtesy of Patrick Hughes.

lucky he dipped his considerably talented toe into its confused waters. Thermodynamics, which expresses conversions between heat and work, emerged in the 1860s only after scientists became convinced that heat was a form of motion. Though Galileo, Descartes, Newton, and others believed that heat was a motion of internal, invisible parts of an object, they didn't have a good atomic theory of matter to back up this idea. Instead, many scientists in the 17th and 18th centuries believed that heat was an invisible fluid, called caloric, that moved freely between substances. If you touched an iron pot fresh out of the hearth, caloric passed from the hot metal into your fingers, causing the unbearable pain of burning heat.

In Espy's time caloric theory still had the upper hand over other heat theories. Since caloric was a substance, it couldn't adequately explain motion, like winds or updrafts. Caloric was adequate to help Espy show that latent heat changed the rate at which saturated air cooled, but the new laws of thermodynamics did much more for storm theory. They enabled scientists to calculate how the atmosphere cooked up enough energy to drive its winds into a storm. By construing heat as motion, scientists could consider heat as kinetic energy—one of two basic forms of energy. In the atmosphere kinetic energy manifests itself as wind and heat. The other form of energy is potential energy, which is basically the potential for motion. Latent heat is such a potential energy; when it is released it becomes the kinetic energy that makes sensible heat. In addition, gravity sets up another type of potential energy in the atmosphere—the potential of altitude. A dense parcel of air high in the atmosphere will fall because of gravity. It may be motionless initially, but it quickly packs plenty of kinetic energy as it becomes a swift downdraft. More generally, potential energy in the atmosphere is manifested by the position of air relative to any apparent "force"—hot just gravity but the buoyancy that drives air up and the pressure gradients that accelerate air horizontally, toward the center of a storm.

In 1862, Lord Kelvin (otherwise known as William Thomson) used the first law of thermodynamics—conservation of energy—to track the gains and losses of kinetic and potential energy attendant in atmospheric convection, making a sort of energy balance sheet. Kelvin's investigation into the thermodynamics behind storms quickly led meteorologists to a breakthrough. Espy believed that moist air might rise unfettered in the atmosphere. But in reality a parcel of moist air will rise only if its environment allows it. Espy had ignored what Kelvin realized: the lapse rate of the air around the parcel tells scientists what potential energy the parcel has and thus how far it will rise. In other words, a parcel is only light *relative to the air around it.*

Meteorologists now knew they had to measure the lapse rate of the air before assessing the potential for updrafts. If temperatures lapse quickly with height, then a moist parcel of air is likely to rise easily. But if temperatures decrease little with height, or even increase (an inversion of the usual situation), then that moist parcel probably can't rise. Clouds won't form, and in the case of a low inversion, all you might get is thick, ground-hugging smog.

The new adiabatic thinking also helped clear up one of the nagging disputes in European meteorology at the time, about a wind called the *foehn.* This infamous dry, warm wind of Alpine valleys can encourage forest fires and sometimes gets blamed for rashes of bizarre crimes and general discomfort. Most meteorologists believed the foehn was air rushing into Europe from the Sahara. If wind is warm, they thought, it must come from a hot place. But in 1866, the Austrian meteorologist Julius Hann showed that in fact the foehn is a local effect. As air rises over one slope of the mountains, it cools and the moisture in it condenses. Further ascent during this cloudmaking stage wrings the moisture out of the air, and the latent heat release slows the decrease in temperature. Then, in the lee of the mountains, the air begins to fall. The now-dry air warms more quickly by compression than it cooled by expansion. At low altitudes this air turns hot, dry, and very bothersome. The snow-eating chinook winds of the Rockies and the hot Santa Anas of Southern California work the same way. Three decades after Espy first brought adiabatic analysis into the science, Hann's explanation surprised many meteorologists who still believed that upper air always carried a chill to the surface.

The rise and fall of air in a storm, like the foehn, responds well to energy balance accounting. Beginning in the late 1860s, meteorologists began using their new thermodynamic tools to account for all of the energy in storms. They were testing themselves: as the air circulated, potential energy became kinetic energy, latent heat became sensible heat, and back and forth. Could the storm theory account for all of these changes? Was Espy's latent heat release during cloud formation enough to drive the whole storm? This new meteorology was a complete departure from men like Redfield and Dove (who finally retired in 1879). Dove's theory that opposing winds caused storms went by the wayside as calculations showed there wasn't enough energy available to make a storm this way. This assessment would later be revised somewhat when temperature contrasts were studied more carefully. For the time being, however, 19th-century theorists were looking at storms as a shuffling of atmospheric energy, which is basically the way meteorology still works, though with many other layers of discovery atop it. At first, for convenience, energy theorists considered the idealized forms of storms most conducive to these studies: Ferrel and Espy provided such a simple model with highly symmetric wind, temperature, and pressure configurations. To many theorists, it didn't matter that synoptic maps were already showing a different story, one that Loomis and FitzRoy, even Dove, in different ways, had tried to tell. Meanwhile, advances in forecasting suffered, because many theorists turned away from the complicated reality of storms.

Hooked on their new, powerful analytical techniques, many storm scientists continued treating all storms as the result of the same circumstances. They saw every violent wind as a part of a closed, pseudo-circular heat machine. The theorists, anxious to flex their thermodynamic muscle, had been studying hurricanes, the ultimate symmetric storm, with great success. But at higher latitudes, where most scientists lived, a different approach was needed, one that ironically brought Dove's belief in clashing winds back into vogue, albeit in updated form. When theory and synoptic experience finally converged, a new way to look at big blizzards and other midlatitude storms emerged, one good enough, say experienced contemporary storm researchers Chester and Harriet Newton, to be a "keystone for modern meteorology."

CHAPTER FIVE

The Aviation Age

Lieutenant Commander Zachary Lansdowne was a nervous man when he gave the orders for the majestic dirigible *Shenandoah* to depart Lakehurst, New Jersey, on September 2, 1925. Before his Navy crew shoved off from their mooring mast at 4 p.m., Lansdowne received a final weather briefing. Forecasters said that skies looked comfortably calm for the first leg of the voyage west, to Illinois. But they also told Lansdowne that before the *Shenandoah* reached the Midwest for its tour of state fairs, a low-pressure area would pass far to the north of the *Shenandoah*.

Lansdowne knew the risks of taking the big airship for a cruise over the nation's heartland. He was sure the low passing to the north meant trouble—thunderstorm trouble. In that region, an impenetrable squall line could stretch hundreds of miles from the center of a traveling cyclone. A native of Ohio, Lansdowne was familiar with the fury of Midwestern thunderstorms. "He had been trying for a year to avoid this flight," his widow later lamented.

With a top speed of barely 70 m.p.h., the *Shenandoah* wasn't built to outrun a Midwestern squall line. Nor was the sheer size of the 680-foot *Shenandoah* comfort in such situations. Dirigibles are large, but they are large so that they can be light, filled with more helium. The *Shenandoah* was the first American-built dirigible, completed in 1923 with German design leadership. The airship held 2 million cubic feet of helium, stored in 20 giant bags, and was notoriously unstable in the winds. The cotton fabric, wire, and cord netting stretched over *Shenandoah*'s skeletal girders were little protection as well. You couldn't even walk a dirigible of that size out of its hangar safely if winds exceeded 5 m.p.h., and you couldn't moor one to a mast if winds were greater than 30 m.p.h. In 1924, high winds ripped the *Shenandoah* from its mooring in Lakehurst and dragged it for eight hours before a crew could gather it in. The accident shredded the nose section, forcing extensive repairs. Lansdowne and his 40-man crew were well aware that they were at the mercy of the elements in their featherweight giant. Somehow they had managed to take the *Shenandoah* on a 19-day cruise of 9,000 miles in 1924.

Early in the morning of September 3, the *Shenandoah* approached Lansdowne's home state of Ohio. In Greenville, not far from Dayton, the commander's elderly mother and many others arose early to watch the *Shenandoah* pass by. They expected their native son to circle overhead as a salute to the town. But the *Shenandoah* never made it.

The dirigible encountered stiff headwinds as it crossed the West Virginia panhandle into Ohio. Storms woke people in Wheeling at around 2 a.m., when the thunder shook their houses. A few looked up at the dark clouds and saw the ghostly profile of the airship silhouetted by lightning. The *Shenandoah* headed west at a comfortable 3,000-foot altitude, but the danger loomed in the sky. Lansdowne maneuvered his ship around the storms beside and ahead of him, ordering at least a dozen course changes in the *Shenandoah*'s last hours.

The crew could not dodge the line of thunderstorms approaching from the northwest, however. At 5:30 a.m., 70-m.p.h. winds brought the great ship to a helpless standstill north of Marietta. The engines strained mightily with their propellers, but to no avail. Then a sudden updraft took the dirigible up to about 4,500 feet. Lansdowne had enough time to order a dive: the nose pointed down nearly 20 degrees with the engines on full thrust. The men frantically tried to release helium to let the ship sink. Nonetheless, another updraft hurled the *Shenandoah* up to 7,000 feet. The blast of air broke the ship's back. The struts carrying the control cabin below the superstructure snapped, sending Lansdowne and 12 of his men plunging to their untimely deaths. The hull split into two; the crew clutched the rigging as each half began to sink. One man, dangling from the ropes, lost his grip and died, but 27 others miraculously landed with only a few injuries. The nose section landed seven miles from Lansdowne's cabin.

After the crash, blame shot back and forth across Capitol Hill as a committee of experts delved into the causes. Some engineers couldn't believe a storm could break the *Shenandoah* in two. They said it disintegrated when it slammed into a hill. A German dirigible expert who helped design the *Shenandoah* said the crew couldn't deflate her fast enough because the Navy had eliminated important safety valves. A British meteorologist blamed the Weather Bureau for providing insufficient weather reports from the Ohio region. The stunned Lakehurst commander attributed the crash to an unforeseeable calamity, like a tornado.

The dirigible era didn't last long. Among other reasons, the inadequacy of weather forecasts deflated the American enthusiasm for airships. In 1933, another Navy dirigible, the *Akron*, crashed in a thunderstorm off the New Jersey shore. Even the infamous *Hindenburg* fell victim to thunderstorms, indirectly. Nearby storms charged the wet, fabric skin of the dirigible as it approached Lakehurst in 1937. When the crew dropped the soaked mooring lines, a spark ignited a hydrogen leak and the *Hindenburg* went down in flames. The common thread in dirigible accidents was not hydrogen, but meteorologists' appalling inability to foresee dangerous storms.

The huge dirigibles were overly sensitive to weather; a simple rain shower could make them ponderously heavy. By contrast, the rickety crates then called airplanes fared better, if not well. If an airplane pilot saw trouble, he could set his craft down in a cornfield. Such emergencies happened all too often at first, but the first aviators trying to keep a schedule—airmail pilots—were desperate to fly no matter what the risks. The pilots took off on orders from Washington, and they only got paid if they flew. No one consulted a meteorologist. One pilot took off for Long Island in 1919 in a winter storm. Snow and 50-m.p.h. winds battered his face in the open-cockpit biplane. With steadily worsening visibility, he barely avoided crashing into the Woolworth building in Manhattan. After restarting his small, cold engine three times in midair, the pilot eventually had to crash-land at Great Neck and transfer the mail bags to the local post office there. His boss in Washington graciously allowed him to take the train home.

A week later, fog blanketed Staten Island as the legendary mail pilot Hamilton Lee desperately searched for the airfield. Lee flew in the shallow layer of clear air just above the choppy waves of New York Harbor. Occasionally he had to dart back up into the pea soup to avoid colliding with a ship. Inside the clouds, Lee could not tell up from down, north from south. Eventually, running out of gas, he took the plunge and landed in the first field he saw. Trees ripped through the fabric bottom of the plane and tore Lee's trousers. The intrepid pilot was back up in the air the next day but admitted, "I shook all the way to Philadelphia."

In the first year, air mail pilots made forced landings every one out of five flights. Inclement weather caused three-quarters of these emergencies. Eventually the pilots organized and struck a deal with the Post Office: a postal manager at the airport would decide if the weather was safe for flying. If the pilot disagreed, he could take the manager along for a test flight. This gave postal workers a chance to test their bravery, too.

Regular airmail service and, later, passenger airlines required better

The 600-foot dirigible *Los Angeles* does handstands on a windy day at its mooring. Lieutenant Francis Reichelderfer, instrumental in modernizing American meteorology, served a tour of duty as in-flight meteorologist aboard this airship. Photo courtesy Patrick Hughes/NOAA.

and better forecasts. Increasingly, pilots turned to the Weather Bureau or to their own private meteorologists, but with little satisfaction. By the 1930s pilots were flying into blizzards and over storms in an effort to keep schedules. The ride wasn't always comfortable for the few passengers. If the plane took too long to hurdle a growing cloud, everyone aboard risked blacking out from lack of oxygen. Flying through storms was no better. One passenger remembered an early Boeing "buck like a wild Brahma steer" in a 1932 Sierra blizzard as the pilot tried to find an emergency airfield. With the throttle frozen at cruising speed, the pilot landed the plane by turning the engine on and off to reduce speed. Eventually the all-metal, pressurized DC-3s dominated airline fleets because they offered economical speed and size and the strength to withstand the punishment of unexpected thunderstorms.

Pilots didn't need only storm warnings from their frustrated meteorologists. They wanted icing and cloud cover forecasts too. Early pilots navigated by landmarks or by the stars, so both low clouds and high overcast layers worried them. If clouds obscured the ground, pilots often flew along railroad tracks—risking a head-on collision with an unexpected tunnel or cowcatcher. At first the Weather Bureau responded feebly to the aviation age. As late as the 1930s, the agency often located its observers in downtown offices even though the airports were miles away, built on converted farmland. Worst of all, forecasting techniques were inadequate. Pushed by disgruntled aviators, meteorology took off as never before in the 20th century, soaring to a new understanding of day-to-day weather, including storms.

One man saw to it that forecasting in America rose to meet aviation standards. He was "the best damn meteorologist in the world," according to his boss, Admiral William Moffett, head of the Navy airship fleet. Francis W. Reichelderfer, born in 1895 in Harlan, Indiana, studied chemistry at Northwestern. He caught the aviation bug in time for World War I, joining the Navy pilot training program, then switched to meteorology to get to the front faster. The British refused his assignment overseas—they were wary of German-surnamed Americans—but Reichelderfer stayed with the Navy when the war ended. He served as a forecaster for Alcock and Brown's historic first flight across the Atlantic in 1919. And when renegade Army Air Force officer Billy Mitchell wanted to show the Navy brass that his bomber pilots could sink a battleship from the air, he turned to Reichelderfer for forecasts. The young Navy officer went along as a bombardier on one of the flying boats. After Mitchell's crews destroyed the surplus German warship *Ostfriesland*, the fliers headed back to Hampton Roads, Virginia, only to encounter a squall line that Reichelderfer had not foreseen. Reichelderfer's plane made it through, but Mitchell and others were lucky to land alive on beaches and coastal farms. "That put the fear of God into me," Reichelderfer later said. "It convinced me that we meteorologists had to do a lot more to help pilots avoid such risks."

Reichelderfer soon hit upon a solution. After the war, Reichelderfer tested his forecast skills by racing balloons. (In 1923, he soared 20,000 feet into fast upper-level winds to place second in the big national competition in Indianapolis.) Reichelderfer realized quickly that he had to ditch the

Early in the aviation age, German pilots used this photo—part of a training guide called "Clouds in the Sea of Air"—to familiarize themselves with the new world above the Earth. This deck of altocumulus would have made navigation difficult in the days before instrument guidance. NOAA photo.

prevalent storm theory with which he blew the forecast for Mitchell's flights. This traditional theory, an 1885 cyclone model devised by Ralph Abercromby, descended directly from Espy and Ferrel. Abercromby said all storms were a pseudosymmetrical spiral of winds around a low-pressure core. With hundreds of maps and observations at hand, this British forecasting guru had assembled a handy snapshot of all the essential ingredients of a midlatitude low-pressure area: a vanguard of high clouds; stormy, rainy areas nearer the center; and clearing skies with cold temperatures behind the low.

Abercromby's model ruled meteorology for more than three decades, but Reichelderfer and others grew disenchanted with it. (One meteorologist later sneered at Abercromby as the "Aristotle" of early-20th-century meteorology.) The model said nothing convincing about the development or motion of a storm—the most important questions for forecasters. Also, Abercromby threw the squall line into his diagram but said nothing about the wind shifts it created or its role in the storm. No wonder Mitchell's bombers got blindsided by a squall line. "Analysis of the weather map was more an art than a science," Reichelderfer later said of the Abercromby days. "The individual artist ... could take great pride in the picture he created, at least until 'tomorrow' came and the weather proved him mistaken."

Jerome Namias, a young man eager to learn meteorology in 1930, also saw how inadequate storm forecasting was. Though he did not have a college education, Namias eventually became a leading weather researcher. He learned enough through his own reading and correspondence courses to obtain a job gathering weather data from archives in Washington, D.C. This gave him ample time to study at the Weather Bureau office downtown and meet the forecasters. Their methods bewildered him. A different synoptic map represented each weather factor—temperature, pressure, wind, humidity. Somehow the chief forecaster mentally pieced all of the information together. Each situation seemed to follow its own rules for where lows would move and how winds would change. The guidebooks were rife with contradictions. "[T]here are hundreds of rules that I tried to memorize in order to forecast," Namias wrote later. "I soon had to give up."

Likewise, forecasting in Britain, according to William Napier-Shaw, head of the Met office from 1900 to 1919, was "partly formulated and partly exercised by the subconscious induction of expertise." In other words it was mumbo-jumbo. The forecasters examined the isobars and somehow extrapolated where pressure anomalies would be in a day. Then they used Abercromby's model to fill in the rain, snow, wind, and temperatures. Forecasters were only as good as their memory of storms and isobars gone by. Not surprisingly, some legendary figures of the period relied on prodigious memory. One of the best forecasters of the era was Henry Harrison, who eventually became head meteorologist at United Airlines. Not surprisingly, Harrison was also an expert bridge player and could rattle off all the latest baseball statistics just as easily as data from the previous week's weather maps.

Reichelderfer was the first American to abandon old-fashioned forecasting methods. He turned to a recent paper by two young meteorologists who studied under the Norwegian physicist Vilhelm Bjerknes. In their storm model, squall lines were integral to the structure of extratropical cyclones. Instead of drawing only isobars on their maps, they analyzed wind and temperature distribution and drew boundaries called fronts between masses of air distinguished by contrasting temperature. Reichelderfer began drawing these fronts for his forecasts at Hampton Roads and distributed papers by Bjerknes' students to his own cadets at the Navy's weather training program. Through Francis Reichelderfer, the meteorology of Vilhelm Bjerknes and his son Jacob gained a foothold in the country where aviation was growing fastest.

Vilhelm Bjerknes did not set out to remake meteorology. A student of Heinrich Hertz at the University of Bonn, Bjerknes at first intended to carry on the work of his father, the physicist Carl Bjerknes, who tried to demonstrate that Newton's laws govern all phenomena, including electromagnetism. Isolated in Norway, a scientific backwater, the elder Bjerknes was quickly losing hope that his life's work would amount to anything. He and his son had few sympathetic colleagues. They represented a dying breed of Newtonian physicists in a dawning age of subatomic particles and relativity theory.

While Vilhelm tried to find analogies between fluid and electromagnetic waves, he stumbled onto yet another seemingly hopeless line of inquiry: weather prediction. While teaching in Stockholm in 1897, Bjerknes discovered a new way to describe the whirling tendencies of fluid flow. He showed that vortices will form in fluid flow because of density gradations that are unrelated to pressure gradations. Where such special density gradations exist, whirls not only form but also tend to grow. In fluids where pressure completely determines density variations, on the other hand, vortices tend to die out quickly.

This might seem like a minor point, a curiosity at best. In fact, a Polish scientist made the same discovery a year before Bjerknes but had dismissed it as unimportant. His reaction is perfectly understandable: we intuitively expect density and pressure to be related. Imagine diving into a swimming pool full of a liquid just like water, only more easily compressed. As you plunge toward the bottom, the pressure increases—your ears might begin to hurt, just like in ordinary water. Swimming in the compressed depths of this pool is very hard work, however. Unlike real water, with higher pressure the liquid in this pool is denser and difficult to pull or kick aside. In this pool, each stroke gets easier the closer to the surface you swim. Density increases as a function of pressure. All is well in this imaginary pool. According to Bjerknes' theorem, vortices won't last long in this liquid. Any swirls along the edge of a current in this pool will quickly dissipate. In the "pool" that is our atmosphere, however, you might suddenly swim into a relatively dense patch of the fluid right near the surface, at low pressure, or a thin area near the bottom. Because of this peculiarity, little whirls can grow bigger and bigger, making storms.

Bjerknes realized that in nature density gradations deviate from pressure gradations. Indeed, geophysics was an ideal field for the Bjerknes circulation theorem. Better yet, though Scandinavia was a physics backwater, it was prime territory for geophysicists, according to Robert Marc

Jacob Bjerknes at the weather chart analysis desk, c. 1920. Photo courtesy of the American Meteorological Society.

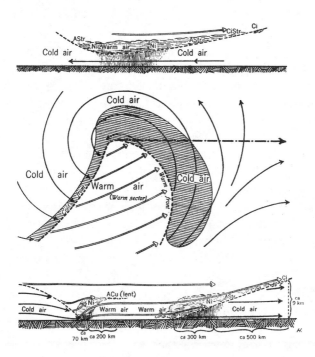

An early diagram of the Norwegian model of extratropical cyclones now familiar on daily-newspaper weather maps around the world. It took decades before experienced forecasters began using the improved model regularly.

Friedman, who has chronicled the career of Bjerknes. Though Bjerknes had few peers in the small Swedish physics community, he knew many leading oceanographers and meteorologists. They quickly realized the importance of the new circulation theorem. One of Bjerknes' friends, meteorologist Nils Ekholm, had been frustrated making forecasts for Swedish balloon expeditions to the North Pole. In 1896 an expedition was canceled owing to bad weather. The next year the balloonists disappeared over the Arctic. Rescue was impossible, because scientists could not reconstruct where the polar winds had taken the balloonists. Ekholm believed Bjerknes had uncovered a way to formulate the stormy paths of the wind.

The new theorem excited Ekholm because it confirmed his suspicions based on extensive study of storms in Norway. He knew that in the atmosphere temperature distribution caused density variations not apparent in the pressure patterns. Since storms redistribute heat, Ekholm concluded that synoptic maps that show only pressure distribution neglect essential quirks in air density. Ekholm realized that Bjerknes' theorem justified his concern, and at the same time gave storm theorists a way to work around the limitations of the synoptic maps that had been the basis of meteorology for a century. The theorem showed how atmospheric density peculiarities generate storms. In 1898 Bjerknes lectured in Stockholm about possible links between his theorem and storms. Ekholm knew he had found a valuable ally and decided to recruit Bjerknes into the field of meteorology. The two sat down together and, journal by journal, read over the latest research on storms.

Ekholm's one-day crash course in meteorology convinced Bjerknes that a good set of upper-air data from kites or balloons would help prove the utility of his circulation theory. Cleveland Abbe and Julius Hann agreed and soon published articles by Bjerknes in meteorological journals in the United States and Austria. In Stockholm, enthusiastic geophysicists bought Bjerknes new kites and instruments to pursue this new avenue of research. King Oscar II gave the distinguished professor permission to conduct kite experiments on royal property. Bjerknes realized that his friends were trying to drag him into a "meteorological vortex." But the contentious field of storm theory was in many ways the perfect arena for Bjerknes. His rigorous physics background would be a great advantage. And he already had a workable theory.

Abbe soon sent Bjerknes kite data from Blue Hill Observatory, near Boston. Bjerknes' new assistant, a bright, former millhand named Johannes Sandstrom, used the data to profile the atmosphere during the

A dust storm rides a cold front across an airbase in Edmonton, Alberta. Pat McCarthy/*Weatherwise* photo contest.

A cold front made visible by cloud and dust moves across Boulder, Colorado. Lynette Rummel/*Weatherwise* photo contest.

passage of a storm over Boston. Then Sandstrom used a graphical short cut (a typical meteorological ploy) to apply the theorem. He sliced the atmosphere with nearly horizontal surfaces of equal pressure and with similar surfaces of equal density. In our imaginary pool, such surfaces would be parallel to each other and to the Earth's surface—but not in the atmosphere. Just as Bjerknes and Ekholm had predicted, pressure and density varied differently—Sandstrom's imaginary surfaces were not quite parallel with one another. The equal-density surfaces crisscrossed the equal-pressure surfaces, forming a honeycomb of tubes. According to the Bjerknes theorem, each tube represented the potential for vortices to form in the atmosphere, so the more tubes, the stronger the atmospheric circulation. Sandstrom counted up the tubes and showed that the pressure/density differences generated the motion of air in the storm.

Sandstrom went on to found modern oceanography with the theorem. He wasn't the only young Bjerknes student to make a mark in oceanography. The circulation theorem was a hit with Scandinavian oceanographers who at the time were puzzling over shifts in sea currents that had devastated Sweden's herring catch in the 19th century. Bjerknes' theorem gave the scientists a rigorous way to explain current movements. Oceans pose a fluid flow problem similar to the atmosphere, because the density of ocean water, like that of air, is due to factors other than pressure—particularly salinity and temperature. The famed polar explorer Fridtjof Nansen asked Bjerknes to help explain pools of calm water that unexpectedly trap ships in otherwise navigable fjords. Bjerknes turned to his student Walfrid Ekman, who used the circulation theorem to show how the pools might form in between currents of different density.

These geophysical ventures were risky for Bjerknes. A colleague in Berlin warned him, "A physicist who goes into meteorology is lost." But times were changing. New upper-air observations enabled meteorologists to test storm theories. At Blue Hill Observatory in the 1890s, Alexander Rotch developed box kites that could carry meteorological instruments for hours. He used miles of piano wire to fly the kites to 15,000 feet, then reeled them back in to analyze the data. Kite research had its dangers, though. One day in 1902, for instance, a kite broke free from the winch of prominent upper-air researcher Teisserenc de Bort of France. The instruments and several miles of steel wire swept across the countryside, stopping a train, knocking down telegraph wires, tying up a gendarme in a park, and finally shackling a ship's propeller in the Seine.

In 1902, Prussian scientists announced a new, improved balloon made of rubberized fabric. It was cheaper than the paper balloons then in vogue and could be calibrated to rise steadily for reliable pressure profiles of the atmosphere to very high altitudes. Within a few years Prussian meteorologists were observing upper winds to guide pilots of the new airships.

In 1904, a year after his father died, Vilhelm Bjerknes plunged into meteorology wholeheartedly. It was time, he declared, to make a lasting mark on science. He told Nansen that his goal was no less than to figure out how to forecast the weather objectively. Two years later, with help from the admiring Cleveland Abbe, Bjerknes secured funding from the Carnegie Foundation—support that continued for 35 years. Bjerknes and Sandstrom set out to overhaul meteorology with rigorous physical analysis. They started a new textbook for the field, using analyses based on the circulation theorem as criteria for revamping the charts derived from weather observations. Bjerknes knew that many balloon observations were useless for quantitative meteorology. For instance, upper-air charts consistently showed a cyclone over Strasbourg, where scientists calibrated barometers differently.

The new meteorology based on the circulation theorem would need to keep track of the movement of air by mass. Bjerknes and Sandstrom saw that this was impossible with the current synoptic maps, which showed surface pressure distribution. So they advocated maps of streamlines— arrows showing wind direction and speed. Bjerknes showed skeptics how streamlines could advance the aviation age. A Zeppelin traveling from Dodge City, Kansas, to Chicago, for example, could save two hours if it followed streamlines instead of going "as the crow flies."

Streamlines proved useful for locating another emerging meteorological concern of the aviation age: squall lines. Scientists were only then starting to publish papers on these lines of thunderstorms. In 1913, Bjerknes moved to the Leipzig Geophysical Institute, where interest in squall lines was acute. Soon after his arrival, the German Navy's first Zeppelin crashed in a squall line over the North Sea. Fourteen men died as the dirigible sank quickly in the frigid water. A month later, another new Zeppelin exploded after take-off, killing 25 more people. The tragedies shocked all of Germany, but Bjerknes quickly seized the opportunity to demonstrate his new meteorology. One of his students in Leipzig used streamlines to show that air converges along squall lines.

In Leipzig, Bjerknes' 18-year-old son, Jacob, began studying meteorology, too. His work on streamlines soon became expert. Vilhelm wrote to a friend, "For the first time, we have made headway with meteorological prognosis based on dynamic principles." But the bloody Great War took its toll on their work. Food was scarce in Germany, and it was hard to concentrate on work. In those bleak days in Leipzig, one of Bjerknes' students managed only one meal a day—a hard boiled egg in the evening. Often that paltry dinner was all he could think about.

Vilhelm's Norwegian friends and relatives worried about conditions in Leipzig. A group of them, including his aged mother, tried to find him a job in neutral Norway. A new geophysical research center at the Bergen Museum opened its doors to Bjerknes. They were not merely doing Bjerknes a favor; meteorology emerged as a critical science in wartime. The importance of weather in war was one of the surprises of the conflict, especially for the Allies. "British soldiers...do not go into action carrying umbrellas," they sniffed in 1914. Such attitudes soon proved deadly. On the front, soldiers nervously watched wind shifts to anticipate poison gas attacks. And daring pilots also needed special forecasts. Unlike mail carriers, military pilots flew behind enemy lines, where you couldn't land safely to avoid a storm. German commanders, for their part, couldn't

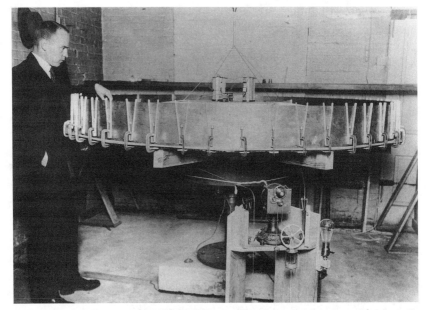

At the Weather Bureau in 1927, Carl-Gustav Rossby experiments with a rotating tank of water that simulates hemispheric motion. Such studies eventually contributed to his theories of the horizontal waves high in the atmosphere that influence storm movement and development. Photo courtesy of Patrick Hughes.

figure out why winds carried Zeppelins far away from their bombing targets in England.

Norway was neutral, but it also desperately needed better meteorology. Poor grain crops and frequent U-boat attacks on commercial ships depleted the nation's food stores. Rationing began in 1917. Meanwhile, unwilling to risk giving the enemy any advantage, England and Iceland had blacked out reports of storms approaching from the west. Farmers and fishermen along Norway's western coast depended heavily on such information. Bjerknes decided to drop his purely theoretical work and focus on forecasting. He proposed a new weather forecast service based at the Bergen Museum. It was a golden opportunity to learn more about storms, even if it meant suspending work with the circulation theorem. He moved back to Norway in late 1917 to begin forecasting the following summer.

Bjerknes immediately began augmenting the Norwegian observational network. The inspiration for his strategy was the German army. Cut off from weather information west of the front, the Kaiser's battlefield meteorologists found they could compensate by making many observations relatively close together. Now Bjerknes did the same thing, increasing by 10-fold the number of observers in western Norway. He recruited farmers and fishermen as well as Navy submarine watchers on the islands offshore. The trained Navy personnel gave him wind direction readings to within 5 degrees, rather than the usual eight compass points (45 degrees apart). This allowed streamline analysis. All the observers telegraphed their reports to Bergen by 8 a.m. each day during the summer of 1918. Jacob and other young assistants used them to make a forecast in about two hours. They wired their predictions to coastal towns where farmers could obtain them with a local telephone call.

Bergen proved to be an ideal setting for hard work and creative thinking. Vilhelm Bjerknes brought together a hand-picked team of bright young scientists to get a fresh start in meteorology. In addition to his son Jacob Bjerknes, Halvor Solberg was one of the stars, a valuable theoretician. In late 1918 the Bergen team recruited recent graduates from Sweden. They came home with Tor Bergeron and Carl-Gustav Rossby. Given the future impact of these two men, Bjerknes' talent haul was the meteorological equivalent of a baseball scout signing young Babe Ruth and Ty Cobb at the same Little League game. Bergeron was an "elderly" 27 and was a junior forecaster in Sweden, but Rossby, barely 20 (a year younger than Jacob Bjerknes), had completed his bachelor's degree in little over a year in Stockholm.

The young forecasters lived with the Bjerknes family and worked together in a large attic room. At lunch they would gather around the table with Vilhelm Bjerknes and talk about the day's forecast and analysis, then head back upstairs for research. "One often sat up the whole night, alone or together with several other young meteorologists ... and pondered over or discussed unexpected developments ...," Bergeron later wrote. One young man apparently didn't leave the house for a week. The few diversions included taking hikes in the mountains around Bergen and attending occasional concerts, where Bergeron and Rossby indulged in a shared passion for music.

Vilhelm Bjerknes meanwhile worked on theoretical problems and logistical issues. Occasionally he would look in on the forecasters, "eyes flashing in anticipation," and ask, "Are there any new discoveries this evening?" The Bergen group was the very opposite of the stifling isolation that had helped destroy the scientific ambitions of Vilhelm's father. The weather was also cooperative. Vilhelm Bjerknes wrote to a Swedish scientist in 1918, "Jack really enjoys his work here.... Steering-lines and squall lines sweep past incessantly and provide him with the best material he could desire for his work."

Streamlines proved to be a powerful tool. In Leipzig Jacob Bjerknes noticed the strong correlation between precipitation and lines of convergence, where opposing streamlines meet and merge. In the first summer in Bergen, in 1918, the young scientists quickly concluded that nearly every storm had a line of convergence that pointed northeast, where the storm was likely to move. On August 14, Jacob saw from his map that a storm would soon hit the west coast. He thought it would move south, and said so in the forecast. The next day, though, the storm moved northeast, following the line where warm airstreams had converged in the eastern part of the storm. He began calling this warm convergence line the "steering line" of the storm.

Jacob Bjerknes and his colleagues saw clearly that extratropical cyclones (storms usually between about 30 and 60 degrees latitude) were not necessarily most intense at the center, contrary to then-current beliefs. Previous storm theory depicted a storm as like a giant fire. The closer to the center, the hotter it got. Similarly, the closer to the center—the energy source of the storm—the stronger the winds. Now the Bergen scientists showed that the intensity concentrated into lines: the squall line had the most significant pressure change and shifting winds. The squall line proved to be the second essential convergence line in the storm, this one running roughly north–south. Together the steering line and the squall line—not the center—also accounted for most of the precipitation. The Bergen team assumed that the lines of convergence met at the low-pressure center. Furthermore, earlier convergence theories predicted that a line of divergence preceded (and paralleled) each convergence line. The convergence and divergence lines formed a symbiotic relationship in the model: air rising immediately ahead of the convergence line curled over and sank, spreading out at the ground at the line of divergence. With this interconnection, the "forerunners" of divergence supposedly ensured the forward movement of the lines of convergence.

Jacob Bjerknes put these ideas together in the fall of 1918 in "On the Structure of Moving Cyclones," an eight-page paper that summarized the new Bergen discoveries. For Francis Reichelderfer, shaken by an unexpected squall line, the paper was a godsend. For others, it came as a shock. Isobars, for instance, were unwelcome in that first summer in Bergen. It wasn't until the group began forecasting winter storms, in 1919, that the young scientists relented and began making maps with the older technique.

In truth, the Bergen breakthrough was really a new interpretation, not so much a revolution. Vilhelm Bjerknes later maintained that the new extratropical cyclone model represented an intense study of observations. But observations quickly got in the way of the original Bergen interpretations. As they drew more and more storm maps, the Bergen team altered their lines of convergence. At first, Jacob Bjerknes depicted air rising ahead and behind the convergence line, as if the air from both sides sloshed together and splashed upward. This upward splash created the recoil that explained the forerunners. But none of the storm observations showed the forerunners, even though theoretically they were supposed to exist. So the Bergen team tossed theoretical formality aside and eliminated the forerunners from the storm model. Now they began to draw the lines of convergence differently, too. The upward splash of colliding airstreams disappeared. In the updated interpretation Jacob showed that one airstream—the colder, denser one—moves under the other.

Even the precipitation distribution in the storm model changed over time. At first rain seemed to fall in a generalized area only slightly ahead of the low-pressure center. This seemed patterned after the slightly asymmetric storm model of Julius Hann published a few decades before. Then, as Bjerknes and others began to account for the effects of Norwegian terrain, they became even more convinced that rain mainly falls along the convergence lines. Friedman concludes from such changes that the Bergen model was a hodge-podge edifice of older theories with observational insights tacked on shingle by shingle. Nonetheless, the Bergen structure soon became integrated into an organic design worthy of Frank Lloyd Wright.

The model grew increasingly more compelling in part because the Bergen scientists began to sense its narrative effectiveness. Bergen team renamed the convergence lines "fronts" in 1919. They took the new, vivid metaphor from the world war. Vilhelm Bjerknes explained that the weather at midlatitudes is an eternal struggle between polar and tropical air along a "Polar Front." Each battle for territory around a low pressure center along the front is a storm: "The warm air is victorious ... east of the center," he wrote, and "rises up over the cold, and approaches ... the pole. The cold air ... presses hard, escapes to the west,... [makes] a short turn toward the south, and attacks the warm air in the flank...." The leading edge of the cold air wedging under the warm sector is called the cold front, and the advancing warm air sliding up over the cold air is called the warm front. The cyclone is a wavy contortion 1,000–2,000 miles long in a grand Polar Front.

At first, the Bergen analysis maps showed the squall line in red and the steering line in blue. Now that they were called fronts and represented the forward progress of cold and warm air, respectively, Rossby wisely suggested switching colors on the maps, making the cold front blue and warm front red. Later, Bergeron sent Jacob Bjerknes a postcard on which he sketched new symbols for the fronts—the black lines with pointed and rounded barbs that you see today in daily newspapers.

Cold fronts and warm fronts are not the same phenomenon in reverse, as one might suspect. Warm fronts move slower and they have such a gradual slope that the warm air rises maybe 50 feet every mile, half the slope of the average cold front. The warm air rises over the cold air ahead of it gradually enough that the clouds are fairly smooth sheets of stratus. The rain is light but long-lasting as the warm front passes. The warm air ascends higher and higher ahead of the front's progress on the ground. A high veil of cloud often announces the approach of a warm front. The sun remains a bright and distinct disk as it shines through this thin, amorphous sheet. Several hours or more later, the overcast is a high-flying, downy blanket—thick enough to obscure the sun, but thin enough to allow plenty of light through. Finally the surface front itself arrives, often more than a day after the first cirrus.

The cold front, on the other hand, strikes suddenly. Temperatures may soon drop precipitously behind or slightly ahead of massive convective towers of cumulonimbus, even severe thunderstorms. That's where the heavy but short-lived rainfall is likely. The sky clears behind the front, and temperatures can drop precipitously. Cold fronts in the American Great Plains, for instance, can send temperatures plummeting from a balmy 80 degrees Fahrenheit to freezing in a few hours.

Without streamlines, the Bergen scientists might not have discovered warm fronts, because warm fronts barely affect the isobars of surface pressure maps. When a cold front passes, on the other hand, pressure is likely to fall significantly, though temporarily. Previous storm theorists had identified cold fronts, or at least squall lines, which sometimes raced ahead of the cold front. But in most earlier storm models, like the Abercromby model, the squall line is a semi-independent feature of the storm. In the Bergen model, the cold front and warm front are both integral to the structure of the storm. They explain the precipitation by forcing warm air to rise, and they are essential to the contrasts between cold and warm air that explain the model.

The first papers to come out of the Bergen school didn't mention similar work from the decades before World War I. This breach of scientific protocol insulted experienced meteorologists and probably delayed acceptance of the Bergen model. Though the young scientists in Norway seemed to be making a clean break from past storm models, in fact their discoveries culminated years of work by meteorologists who step by step broke down the unanimous acceptance of the symmetrical, thermal theory of storms initiated by Espy and later Ferrel. To do so, storm theorists had to shift attention from the release of latent heat at the center of midlatitude storms to the sharp temperature contrasts between the airstreams entering storms.

Frank Bigelow, a prominent theorist with the U.S. Weather Bureau, looked over thousands of observations of American skies in 1896–97 and concluded that air rotates between atmospheric currents moving in opposite directions. He saw three main currents in storms: an upper-level, westerly flow; and, at the surface, a cold, northerly flow and a warm, southerly flow. Furthermore, Bigelow correctly ascertained that "the counterflow and overflow of currents of air of different temperatures" provided the energy for midlatitude storms.

Meteorologists in America didn't accept Bigelow's ideas readily. Americans clung proudly to Ferrel's idea that all storms have a core of rising warm air. But Europeans began to doubt the validity of the symmetrical storm model. Mountaintop data in Austria suggested that the core of a large midlatitude cyclone is actually cooler than its surroundings. Then

Top: Weather Bureau meteorologists launch a pilot balloon to observe winds aloft. Above: Navy planes routinely patrolled the skies for meteorological observations over the Washington, D.C., area before radiosonde balloons allowed safer, remote observations. NOAA photos courtesy of Patrick Hughes.

the Reverend Clement Ley, an amateur observer in England, made careful measurements of cloud motions in the 1870s. He found that the center of a low-pressure area is not vertical but instead leans backward from surface to upper air. This remained a hotly contested observation until the famous German meteorologist Wladimir Koeppen explained a decade later that cold air gathers in the back of the storm. This makes the air near the surface denser behind the low than in front of it. Pressure drops faster with height in denser air, so that, at high altitudes, pressure is lower above a pool of cold air than above a pool of warm air. In a typical eastward-moving storm with a cold backside, the low-pressure center lags farther and farther westward, or backward, with height. Thus the Reverend Ley's observation was a critical piece of evidence favoring the eventual development of the Bergen model. The tilted cyclone center required asymmetry of hot and cold air, exactly as the Norwegians eventually modeled it.

Norwegian scientists before Bjerknes also contributed to the burgeoning evidence of asymmetrical temperature distribution in extratropical cyclones. At the Norwegian Meteorological Institute, Professor Henrik Mohn studied hundreds of synoptic maps for an analysis of storms he published in 1870. To get better resolution, he mapped the pressure and temperature distribution of the atmosphere four times a day. His maps showed that a tongue of warm air extends into a midlatitude storm from the southeast, an early recognition of the "warm sector" between fronts eventually included in the Bergen model. Mohn also correctly pointed out that midlatitude storms form where surface temperature gradients are intense. Such gradients exist near Newfoundland, he said, where cold water from Baffin Bay meets the warm Gulf Stream. The air picks up the local surface contrast, often more than 35 degrees, and breeds many of Europe's worst storms. Storms traveling up the East Coast of the United States similarly get a jolt of energy (or regroup entirely) when they encounter the Gulf Stream off North Carolina. The western Pacific, near Japan, is also a storm breeding zone, as are mild Pacific waters off the cold Alaska coast.

One of the most significant new ideas that Mohn and Bigelow brought to storm theory was that air continually circulates into a low-pressure area. Ferrel, on the other hand, modeled all storms as closed circulation systems, in part for simplicity's sake. He insisted that any energy needs were met within the confines of the low-pressure area. This is fine if you think that the center of the low is where all of the action is; indeed, if convection drives a storm, that's close to the truth. But the new theorists made it possible to fuel a low-pressure area with energy from outside air.

The Austrian Max Margules put a mathematical face on these ideas in the first decade of the 20th century. In an ingenious thought experiment, he asked what would happen if two masses of air—one warmer than the other—collided. Would temperature contrasts create a storm? Margules, a chemist by training, described the conversion of potential to kinetic energy that occurs as cold air and warm air sort themselves out, seeking a proper altitude based on density differences. (As a whole, the center of gravity of a storm sinks as lighter air ascends and colder air descends.) Margules found that temperature contrasts pack enough energy to drive the winds of a storm. Furthermore, Margules declared latent heat release to be a red herring, at least in midlatitudes. Convection-released latent heat accounts for only about one-tenth the energy required to explain the movement in an extratropical cyclone.

By World War I, William Napier-Shaw, head of the British Met Office, also believed that contrasting air clashes in midlatitude storms. He used his synoptic experience to draw a diagram of storm flow that was very similar to the later Bergen model. But ironically, though Shaw was in a position to break the hold of the Abercromby model in England, he did not apply his research to actual forecasts. None of the new-style storm theorists before Bjerknes changed forecasting procedures significantly with their findings. Margules got so fed up with storm studies that he returned to chemistry, only to die of starvation and neglect. The Bergen scientists finally caught the attention of forecasters because they used their theories in day-to-day forecasts and because they went beyond a credible static model of midlatitude storms. They also showed how storms develop and move. This was a major achievement, and it might not have happened without Tor Bergeron's careful analytical eye.

Vilhelm Bjerknes wrote admiringly that Bergeron "is certainly one of the best prognosticians that could be found.... He knows how to exploit every single sign either on the chart or in the sky." Bergeron earned this praise by discovering the "occlusion" stage—the logical culmination of the growth of the cyclone. For weeks in 1919 the young Norwegian had toyed with the idea, especially when the cold front of a heavy snowstorm on November 12 seemed to catch up with the warm front. Finally, on the edge of the November 18 map he sketched a cross section of the storm, showing how, after the cold front catches up with the warm front, the faster wedge of cold air shoves the old warm sector air off the ground. After the occlusion, warm air spreads outward from the low center at midlevels while cold air fills up the center of the storm.

An occluded front barely ripples the surface pressure and precipitation distribution. The old cold front once brashly strode into the warm sector, clearing out the clouds behind it; now it is obscured, with rain behind it as well as at its vanguard. Without a significant moisture or temperature contrast, the cold front has lost its edge. In fact, during an occlusion, the cold air basically is advancing into itself; the head of the cold air is merely biting its own tail. The surface edge of the cold front lacks the punch to make heavy precipitation.

Upon occlusion, the cold front sometimes blunders into air that is colder than itself. In this case, the cold front gets a rude awakening. It is lighter than its new environment, so it actually rises—above the colder surface air, but below the warm sector air lofted above. This creates a sandwich of air that forecasters must probe carefully. The lifted warm sector cools with expansion and forms precipitation—perhaps snow. The crystals may melt as they fall but then they encounter a layer of very cold, dense air hugging the ground. If this layer is not particularly deep, the drops might cool below 32 degrees Fahrenheit but freeze only when they hit the ground, coating trees, power lines, and streets with a weighty glaze of ice. Such freezing rain caused extensive power outages in the South during later stages of the Blizzard of '93. The thick sheet of ice under the snow and slush made the roads treacherous.

The occlusion is a helpful concept because it has a logic of energy behind it. If you think of the cyclone as a wave along the polar front, then you have a pretty good idea of what it looks like through most of its life cycle. It is tempting to think these waves are similar to those on the ocean surface. After all, ocean waves form at the interface between dense water and less dense air, and Bjerknes' circulation theorem says whirls of motion grow at density interfaces. But unlike ocean waves, the storm wave distorts an unstable density interface. The growing amplitude of the wave in the polar front brings warm air uncomfortably close to cold air in a concentrated area. The growing temperature contrast along the wave is the energy source of the storm. It is the atmospheric equivalent to putting Bengal tigers and polar bears together in the same cage. Fur will fly.

Something has to end this ferocious buildup of potential energy, and Bergeron's occlusion does the trick by swamping the center with cold air, which blots out the temperature contrast. The kinetic energy (motion) of the storm peaks, and friction inevitably wears away the storm's winds. The storm even moves slower as a whole, sometimes stalling soon after occlusion.

Tropical clouds band together under a dry but powerfully swift subtropical jet stream, pouring flooding rains in California. Two branches of the clouds detour into skirmishes with polar air in the curling circulation of midlatitude cyclones—one north of Hawaii and another near the West Coast. NOAA satellite image courtesy of Lee Grenci.

A few months after Bergeron discovered the occlusion, Halvor Solberg looked at one of the weather maps and realized that if he traced the cold front of one storm far enough back, it turned out to be the warm front of the next storm to the west. In this way he realized that storms travel in families, holding hands. Solberg's map analysis suggested that the polar front stretches across the entire globe.

A few years later, on his first airplane trip, Bergeron excitedly took notes about the clouds forming around him. Later, looking at the weather maps for the flight, he identified important phases in the evolution described in the Bergen model. During the flight the polar front was a stationary chain of small waves. One of the small undulations took three days to develop into a large cyclone and move across Europe. Bergeron's study was a tour de force of synoptic analysis and powerfully demonstrated not only Solberg's cyclone family concept but also the importance of analyzing developments aloft directly. The Bergen school had used cloud observations as a substitute for regular upper-air measurements. Rossby and Bergeron were convinced this wasn't enough (and Jacob Bjerknes eventually came around to this opinion, too). Ernst Calwagen, another young member of the Bergen team, began flying small planes to gather upper-air data to expand the new storm model. In August 1925, he crashed and died. The incident shook Bergeron, who never set foot in an airplane again.

Despite the effectiveness of the Bergen model, at first forecasters shunned it. Ten years after Reichelderfer began using Norwegian techniques for the Navy, Namias stumbled onto papers from Bjerknes, Solberg, and Bergeron, dutifully tucked into the shelves of the Weather Bureau office in Washington. Practically no one had bothered to look at them.

The tide was turning toward the frontal model, however. In 1926 Reichelderfer gained an energetic new ally in his crusade: Carl-Gustav Rossby. Rossby arrived in Washington that year on a fellowship to promote Scandinavian–American scientific understanding. He went to Washington as an emissary for the Bergen school at the Weather Bureau. With Reichelderfer's help, he ended up staying in the United States for 25 years. His impact on American meteorology was profound. His mark on storm theory was nothing less than revolutionary.

Rossby emerged from Bergen with near superhuman ambition. Some say he was compensating for his own physical frailty. As a boy, he had been weakened by rheumatic fever. But not even Rossby's closest collaborators knew this about him. He was a dynamo, constantly jumping from one project to another. He was a charmer, who inspired dozens of young men and women to tackle projects he dreamed up. And he was an exceptional scientific talent who had an amazing ability to simplify and solve the most complex theoretical problems.

If Rossby had a weakness, it was forecasting. At Bergen he got a reputation for somewhat sloppy, cursory analyses of the weather. But Rossby shared with Vilhelm Bjerknes a dual talent for both theory and leadership. Everywhere he went, he created a fizzing atmosphere of scientific achievement, collaboration, creativity, and theoretical brilliance as well as assiduous attention to practical needs and careful observation. This potential was apparent from the start, Bergeron recalled: "This boy ... had an amazing persuasive and organizing faculty; his far-reaching ideas and high-flying plans often took our breath away." Some think Rossby's experience in Bergen, where he remained in the shadow of greatness, drove him to make his own mark elsewhere.

Rossby came to Washington to introduce Bergen methods to forecasters at the Weather Bureau. The older men tolerated the young Swede but basically brushed him aside. Rossby took the opportunity to pursue research of his own, and in the basement he devised a round rotating tank six feet in diameter to simulate stormy flow in fluids. Meanwhile, he couldn't help but notice the Navy forecaster who visited the Weather Bureau each day, collected the current maps from the other meteorologists, and went to his own corner of the office to quietly make his own analyses, using fronts. Reichelderfer and Rossby struck a crucial friendship.

After two years in Washington, Rossby was no longer welcome at the Weather Bureau. Part of Rossby's problem in Washington was that meteorologists believed that North Sea storms are fundamentally different from American storms. A few early studies made it clear, however, that if anything American fronts are even more pronounced than their European counterparts. Observations were also an issue. One of the first Americans to study with Bjerknes in Bergen, Anne Louise Beck, wrote in *Monthly Weather Review* about the new technique, leading one prominent meteorologist to complain that the Weather Bureau would have to expand from fewer than 300 to nearly 5,000 observing stations to use frontal analysis. Greater detail in wind measurements would also be necessary for streamline analysis. Rossby wisecracked that the old-timers at the Weather Bureau hadn't needed such observations earlier because they made up the forecasts anyway. In any case, Rossby's days in Washington were numbered. He incensed his bosses by going around Weather Bureau channels to make forecasts with Bergen methods for a high-profile flight to South America by Charles Lindbergh.

Reichelderfer came to the rescue, introducing Rossby to one of his ballooning buddies, Harry Guggenheim. Guggenheim, who was a World War I pilot, convinced his father to establish the Daniel Guggenheim Fund for the Promotion of Aeronautics, which among other things funded Lindbergh's celebratory cross-country tour after he crossed the Atlantic. The rocket pioneer Robert Goddard was one of the many scientists supported by Guggenheim funds. Harry Guggenheim gave Rossby money to improve the forecasting for a demonstration airline route in California, and Rossby headed to Oakland in 1928. Though the head of the Weather

Bureau warned his California forecasters about Rossby's impetuousness, the young Swede got a fresh start in California. He hung out with the pilots at the airport and heard their stories about weather conditions along the coast. The Los Angeles to Oakland airway wasn't safe without better forecasts, they told him. So Rossby convinced an Army pilot to fly him from town to town along the route. In those heady days of flying, aviators were considered Lindbergh's brethren. Local officials treated the enthusiastic young meteorologist like a hero and quickly agreed to help him improve weather observations in the state. By spring, Rossby had a new forecasting service of his own at the airport. Observations arrived by a fancy new teletype network, and his own team of forecasters then used Norwegian methods to brief pilots before takeoff. The pilots were impressed, and so were the Guggenheims.

The Guggenheims soon tapped Rossby to head the first dedicated meteorology program at an American university. While Rossby was in California winning pilots over to frontal analysis, Reichelderfer had set up a Bergen-style training course for Navy meteorologists. After a year at Annapolis, the course moved to Cambridge, Massachusetts, courtesy of Harvard and MIT. Jerome Hunsaker, chairman of aeronautical engineering at MIT, like Reichelderfer, was concerned about the fate of the Navy dirigibles. Not only was he a Navy man himself but also he had helped design the ill-fated airships. He knew aviation would get nowhere without better meteorology. With Guggenheim money Hunsaker expanded the Navy courses into a full meteorology program—for starters, within MIT's aeronautical engineering department. Rossby was the man to run it, and at 31 he became an associate professor at MIT, beginning a golden age of meteorological research in the United States. One of his first students, Horace Byers (who had assisted Rossby in Oakland), said, "Those who studied under [Rossby] practically worshipped him. They were participating in his great crusade—to bring modern meteorology to America." Rossby regularly joined his students for lunch or coffee at a cafeteria near the school. There he would inspire them to tackle important problems in oceanography and meteorology. They rarely left the table without a full stomach and a full head of brilliant new ideas to pursue.

Rossby's agenda at MIT was to tie up the loose ends in the Bergen model. The new model shifted research away from low-pressure areas and toward high-pressure areas, called air masses. To understand midlatitude cyclones, meteorologists needed to understand the air masses that clashed. The temperature differences and wind shear between air masses determine the slope of fronts. The temperature and humidity of air masses determine the type and intensity of precipitation at fronts. Rossby brought a German expert in scientific measurement and aviation to MIT to design flight-worthy meteorological instruments for air mass studies. Then Hunsaker offered free classes to the Army Air Forces in exchange for an airplane and a pilot to fly it. One of the first students in the exchange was Arthur Merewether. The lieutenant arose at 5 a.m. every day to fly over Boston—rain, snow, or shine—giving his professors a steady stream of upper-air data.

Rossby was always on the lookout for talent, just as Bjerknes had been. A young New Englander so impressed him with a letter critical of one of his recent papers that Rossby invited him to stop by and chat. Jerome Namias showed up a month later, and Rossby hired him to analyze the profile of air masses with the MIT airplane data. Namias learned to analyze air masses a new way—with potential temperature, the temperature the air would have if moved to a standard sea-level condition. This is called isentropic analysis, and it revealed the graceful motions of air across the surface of the globe. Namias remembers showing some of his isentropic analyses to his wife, who had trained as an artist. "Edith had ... a wonderful feel for symmetry, balance, and aesthetics. She often found that parts of my isentropic analyses jarred on her sensibilities and wanted to amend them.... I soon discovered that those offending parts were also scientifically incorrect, and that Edith's modification was nearer the mark."

Air mass analysis was only a start to improving the Bergen model. Meteorologists desperately needed a better idea of what happens at the top of the troposphere. William Dines, a leader in kite research at the turn of the century, had said that a storm forms at the surface when divergence in the upper atmosphere is sufficient. Upper divergence helps clear out the air and create low pressure. The technology for sustained upper-level research wasn't available to Dines, however. When meteorologists created radiosondes (balloons with radio transmitters) in the 1920s, they could get temperature, pressure, and humidity reports from free air in real time. Now meteorologists could complete the Bergen model with upper divergence. (Today radar also tracks the balloons for additional wind information—meteorologists launch thousands of 'sondes around the world at noon and midnight, Greenwich Mean Time, to maintain the backbone of the world's forecasting services.) To flesh out the storm model, Jacob Bjerknes and the Finn Erik Palmén studied data from special international balloon launch projects. They found that fronts at the surface were indeed very narrow zones of temperature contrast, perhaps 50 miles wide, but that at 10,000 feet and higher, fronts were not as easy to define—often occupying zones of air 100 or more miles wide.

In the United States, a weather problem far more pressing than the growth of aviation made the upper-air flow a hot topic. The Dust Bowl of the 1930s was ruining farmers on the Plains in the middle of the Great Depression. The Agriculture Department turned to Rossby and MIT to improve long-range forecasts to help farmers. Rossby's work on upper westerly flow held great potential for forecasting weather a week in advance. Once again, Rossby put Namias to work. Namias began using sparse upper-air data to show practically, quantitatively, what Rossby was cooking up for meteorology—a new emphasis on giant waves in the upper atmosphere. The waves are north–south undulations in the westerly flow aloft, and they have crests anywhere from 40 to 90 degrees longitude apart, meaning only about four to seven girdle the globe on any given day. Meteorologists call a southward bend of flow a trough, since it represents an area of low pressure aloft. Low-pressure troughs aloft lag behind the surface low pressure of midlatitude storms, just as the Reverend Ley had figured from the movement of cirrus clouds in the 1870s.

Namias drew Rossby's waves on his maps even though he had enough data for only about half a wavelength of the giant circulation feature. Then, scribbling a simple equation of Rossby's in the margin, he showed convincingly that the speed of the wind and the wavelength could tell you how the waves move. The movement of the waves in turn determines how strongly air spins relative to the Earth's surface. The progress of the waves indicates the movement of storms across the midlatitudes. Finally, objective storm forecasting had real promise.

Rossby gave a talk at an international meeting of meteorologists in Toronto entitled "Planetary Flow Patterns in the Atmosphere." It was typical of his sweeping visions of theory. Rossby concluded that the best way to judge where and how fast a disturbance in the atmosphere will move is very simple: Look at the twist of flow, or "vorticity." The vorticity distribution, Rossby showed, was more important than even the pressure distribution. "This was the key to the coming revolution in dynamic meteorology and in forecasting," says George Cressman, a former head of the National Weather Service. It represented an overhaul of meteorology, a "stunning conclusion" in "a paper of unrivaled significance in the history of meteorology."

Cressman had studied the Norwegian techniques at Penn State on the eve of World War II. He had heard critics argue that a model based on North Sea weather was probably just a regional phenomenon, then in the war had seen that model help forecasters predict weather throughout the world. Now, with Rossby's stunning results, Cressman and other scientists spent the next decades investigating the repercussions of Rossby's paper.

Curves within Rossby's waves produce the upper-level divergence required by Dines. Air slows as it enters a trough, but speeds up as it sweeps out of a trough and begins to trace the curve of a ridge of high pressure. The acceleration is one form of divergence—spreading out air along the wave—and accounts for the development of many storms. This upper-level action was the missing ingredient in the early Bergen model. Without divergence aloft to overcompensate for convergence at the surface, there will be no storm. When the large, long-lived Rossby waves develop a small, temporary troughlike kink, meteorologists often see a concentration of rising air underneath. These short waves embedded in the Rossby waves are the truly potent areas of storm formation, drawing air masses together.

While Rossby ran the MIT department, Reichelderfer continued his Norwegian-style forecasts for the Navy. His maps nonetheless often didn't agree with those produced by the MIT analysis team. So Reichelderfer traveled to Bergen in 1933 with a sheaf of synoptic maps and consulted with Jacob Bjerknes. When he came back, his report on air mass analysis of storms was marked "Restricted" by the military brass. This only piqued other meteorologists' interest, and bootleg copies of Reichelderfer's notes spread like wildfire. The pressure for change eventually went all the way to the Commander in Chief. Pressed by the aviation industry for better forecasts, President Franklin Roosevelt ordered a review of the Weather Bureau's forecasting techniques. He set up a board of inquiry including the presidents of MIT and Caltech (which had established its own, modern meteorology department in 1933). The chief of the Weather Bureau, 74-year-old Charles Marvin, knew his time was up and resigned. After the sudden death of his successor, Willis Gregg, in 1938, the review board convinced Reichelderfer to resign his Navy commission and take over the Weather Bureau. Rossby became his assistant for research and training, and the two began to retool the Weather Bureau for Norwegian weather analysis. After a year, the peripatetic Rossby moved on to the University of Chicago to launch another noteworthy chapter in American weather research.

Rossby's Chicago period is celebrated for the discovery of the jet stream. In the early 1920s, meteorologists began to suspect that upper winds reached incredible velocities. One sounding balloon launched in England in 1922 somehow made it to Leipzig, Germany, in four hours—an average speed of 143 m.p.h. Meteorologists believed the balloon probably traveled at 200 m.p.h. at the top of the troposphere. Rossby also suspected that his waves contained concentrated streams of high-speed winds, but he had no good proof of this. Then, in World War II, German pilots noticed how slowly they made progress toward England. Their lumbering bombers seemed to stand still over enemy terrain. American pilots noted the same problem over Japan: they couldn't make headway into fierce winds over the target, then cruised back to their bases at incredible speeds. At first, forecasters didn't believe the pilots' accounts, but it quickly became obvious that narrow bands of high winds—jet streams—coursed through the top of the troposphere.

Rossby's group at the University of Chicago took a close look at the evidence. They made daily upper-air charts and listened closely to daily weather briefings given by George Cressman. In particular, Erik Palmén, a frequent visitor to Chicago from Finland, developed the theory of jet streams from these charts. The bands of winds form in precipitous pressure gradients at high altitudes. In the northern hemisphere, the jet flows eastward with cold, dense polar air below to the left and warm, light air to the right. The density differences are slight at the surface, but high in the atmosphere they are so extreme that the troposphere top is fractured, being much higher over tropical air. A corresponding pressure force pushes upper winds northward, but the Coriolis effect swings the air eastward, creating bands of high-speed winds.

The polar jets, over midlatitudes, are bound to the surface frontal zones, though there can also be a more southerly subtropical jet over regions with no fronts. The polar jets migrate with the seasons, belting the globe at higher and higher latitudes as winter gives way to summer and cold air retreats poleward. The jets, which are about 100 miles wide, are not continuous around the Earth. Rather, the jet stream is really a series of short, straight jet streaks hundreds of miles long. At the entrance and exit of each of these streaks are pairs of diverging and converging flow. The divergences, naturally, make air rise and storms form.

During Rossby's years in Chicago, he also managed to put American meteorology on war footing. He initiated a crash course in weather forecasting for the military at Chicago, Caltech, MIT, UCLA (where he had gotten Jacob Bjerknes and Harald Sverdrup new jobs), and other schools. Rossby's own department at Chicago graduated 1,700 of the 7,000 meteorological officers trained to predict weather for invasions, bombing raids, and naval operations around the globe. The war that turned out to be the first great proving ground of the aviation age was also the first great massed effort in weather forecasting.

In addition to the huge influx of scientific talent, meteorology has benefited immeasurably from the technology of World War II. In particular, the study of storms leapt forward with the addition of radar, satellites, and computers to the meteorological tool kit. As we shall see, the new technology confirmed the value of Rossby's new, top-down approach to midlatitude storm systems. The postwar boom in meteorology has revealed many complications and intriguing twists to the highly idealized model sketched back in 1918 in Norway. In the meantime, however, the Bergen school freed meteorologists from many paralyzing half-truths of 19th-century research and opened the way to discoveries about precipitation, clouds, and other wonders working essential magic within storms. Many meteorologists got so overconfident with their new handle on the atmosphere that they preached not only understanding and predicting storms but also controlling them.

CHAPTER SIX

Brewing the Storm

A storm with a sharp cold front developed in the lee of the Rockies on February 8, 1994. It was a perfect day for cold and warm air to clash over a wintry landscape. To the north, Promise, South Dakota, endured a frigid reading of −52 degrees Fahrenheit. In Texas, closer to the front, Austin was 136 degrees warmer. The storm fed on the temperature disparity and swept through Texas on Wednesday the 9th. In less than a day Dallas cooled almost 60 degrees to a low of 20. But, like most storms, this was a story not just of temperature but also of water. Water in many forms.

That Wednesday, schools in Corpus Christi, Texas, closed because cold rain froze instantly on the roads. The freezing rain mixed with snow in Dallas, where traffic accidents were triple their normal rate. Airlines canceled more than 400 flights out of Dallas–Fort Worth Airport because the water falling steadily from the skies was cold enough to freeze on impact on the wings of low-flying jets.

The atmosphere gave the storm the full red-carpet treatment as it paraded into the East Coast. The conditions groomed a spectacular deluge. The polar jet stream prepared the storm's arrival by ushering cold Canadian air into New England while the subtropical jet invited moist maritime air into the mid-Atlantic region. This jet protocol had hosted a series of winter storms in the East in January and, with them, record cold outbreaks. That month, New Whiteland, Indiana, set a state record with −36 degrees; and Crown Point, New York, on Lake Champlain, had registered −48 degrees. Now Lake Superior was frozen practically from shore to shore for the first time in 16 years; icebreakers worked steadily to free three times as many stranded freighters as usual. But while Texas and the northern states reeled under ice and snow, temperatures soared into the upper 80s in South Carolina. A persistent east–west frontal zone developed over the mid-Atlantic states. Then the storm approached and began to ripple the front. Strange things happened—not tornadoes or high winds, but sudden, frequent fluctuations of precipitation. One moment it snowed quietly, the next moment freezing rain smacked sidewalks, crystallizing immediately into a solid sheet of ice. Sometimes tiny chunks of ice—sleet—bounced into the mix. In North Carolina, the wintry precipitation arrived only hours after the temperature hovered near 70.

Patrick Michaels, a University of Virginia climatologist, marveled at how the atmosphere in Washington, D.C., seemed to mimic the capital city's indecisive political ways. He called the mix of sleet and freezing rain "sleeze." The icy moods of the winter there had already forced some children to miss more than two weeks of school. Three inches of sleet and ice now coated the District, making the narrow, hilly streets and sidewalks of many neighborhoods treacherous. Youngsters skated and played hockey on the reflecting pool under the watchful gaze of the Lincoln Memorial.

Away from the fickle frontal mix, the storm was more consistent. Across the South, from Arkansas and Mississippi to the Carolinas, freezing rain fell pretty much everywhere north of Atlanta. Each pine tree sagged with several tons of ice on its boughs. Several inches of ice glazed power lines, and the weight snapped the stout poles in two. The toppled trees and power poles strewn across the landscape prompted the mayor of Greenville, Mississippi, to compare the scene to war-torn Bosnia. Water refused to flow in many pipes, and electricity failed for two million people. In many states, National Guardsmen wielded chain saws, not weapons, to clear broken branches. But new ice and snow fell relentlessly the following weekend. Utility workers in Baltimore fixed a stretch of downed power poles only to have a tree fall on the line and break it in the new storm.

North of the frontal zone, snow fell twice that week. In one storm, Boston got 18 inches, and in another, New Yorkers got 12 inches. Local airports shut down, and auto body shops boomed with repairs after the latest wave of fender benders. Like most cities in the East, New York had used vast amounts of salt and sand on roads in January. Now with more than a month left in the snow season, the city used a third of its remaining 16,500 tons to keep major avenues open.

Time and time again in the winter of 1993–94, storms worked wintry wonders with precipitation east of the Rockies. Without great violence or remarkable pressures, the storms were mediocre by some meteorological standards. Nonetheless, they wielded a potent combination of temperatures and water doom, causing billions of dollars worth of damage. Many cities set seasonal records for snow, including Boston with 96 inches (a total later topped by a foot in the snowy winter of 1995–96), and Erie, Pennsylvania, with 131. In all, nine separate storms dropped paralyzing forms of the familiar hydrogen-and-oxygen molecule: sleet, freezing rain, snow, or just plain water in torrential amounts.

Water is always overhead, waiting for the right moment to enter our lives. On any given day, the column of perfectly clear, blue sky above has enough water evaporated in it to make a half inch or more of rain—usually more than an inch in temperate climates. A storm summons this genie out of the bottle to perform its magic. Indeed, the Superstorm of '93 was a great storm in part because it was incredibly prolific with water: In less than three days, it delivered as much water out of thin air as flows through the lower Mississippi River in 40 days and 40 nights. That's 55 billion tons of rain, sleet, and snow.

Clearly, to appreciate storms in their full glory, meteorologists must understand where that water comes from and how it falls. This is the science of clouds, a specialty that demonstrates the wide range of scales of agitation in the atmosphere necessary to make a storm. Some scientists like

After the driving, freezing rain of a Rochester, New York, storm, slender tentacles of ice encase the delicacy of detail in the winter landscape. Carl Wetzstein/*Weatherwise* photo contest.

to compare the interaction of the micro- and macrocosms of the atmosphere to the dense structures of a symphony. To understand a symphony in all its complexity, you must understand the relationships between slow and fast movements, as well as between themes within those movements. You must recognize key changes and harmonic shifts. You must get down to the level of instrumentation and phrasing, note by note. The atmosphere is indeed similar. The smallest whirl of leaves on an autumn afternoon ultimately interacts with the giant undulations of jet streams across the planet. But the individual notes of the atmospheric symphony lie in the microscopic world of tiny water droplets.

In a sense, meteorologists began conducting this giant symphony before they could read the notes. With fronts and jets, meteorologists knew the tempos and movements well enough to begin predicting where and when storms might bloom. But they didn't understand raindrops and snowflakes. When they looked up at the clouds at the turn of this century, meteorologists couldn't explain precipitation or even clouds. They didn't understand the mysterious microcosm of water that determines the course of clouds and storms. When meteorologists finally began to grasp the essential workings of atmospheric water about 50 years ago, they got so carried away they began to compose their own weather. The results were not always harmonious, to say the least.

Many early thinkers believed that clouds were collections of particles more like colorful soap bubbles than tiny raindrops. One reason for this idea was that thin clouds produce iridescence when sunlight filters directly through them. The bubble idea persisted for decades into the 19th century, and the discovery of helium and hydrogen led some scientists to speculate that the bubbles contained light gases in order to stay aloft. One early mountain climber took a magnifying glass on an expedition to the Alps and thought he proved the bubble theory by observing cloud particles pop in thin air. But optical physicists proved in the mid-19th century that groups of uniformly sized water droplets can refract light into iridescence. The more uniform a cloud, the brighter the corona. Balloonists, who endured frosted clothes and beards at high altitudes, also realized that clouds are water or ice. In addition, James Espy showed that rising air causes condensation in the atmosphere. In all, the bubble theory died a well-deserved death. But exactly how water condenses remained a mystery. And as scientists began to realize that cloud droplets were tiny enough to stay aloft in gently rising air, they wondered how some drops grow large enough to fall as rain.

Scientists in the late 19th century began studying condensation in earnest and revealed the invisible processes at work. They saw that if you pour water into a glass and let it sit, the surface of the water is neither still nor inactive. The water and the vapor above it wage a little battle in search of an equilibrium. Molecules of vapor diffused in the liquid sporadically rocket out of the water, like missiles launched from submarines. At the same time, vapor molecules in the air plummet like depth charges into the water. If the outgoing gas molecules exceed the incoming air molecules, then we say the liquid is evaporating. In a dry, warm, well-ventilated place, the water in the jar will evaporate completely if given the time.

The vapor escapes from the liquid until the pressure of the vapor just above the surface of the water is high enough to reach a special threshold, called the saturation vapor pressure. At the saturation vapor pressure, the water and the vapor above it are at equilibrium, and the amount of vapor entering the liquid equals the amount of vapor leaving the liquid. The saturation vapor pressure is independent of the pressure of the other air molecules near the surface of the water, but it does depend on tempera-ture. When temperature rises, the saturation vapor pressure increases—meaning more vapor can escape the water. On a hot, humid day, air has much more water than on a cold day.

But this description of water at saturation is all theory—quite accurate when you have a dish of water handy. As early as 1879, scientists realized that high in the atmosphere water doesn't condense simply because the temperature and pressure are right. The vapor needs help. When you serve cold lemonade on a hot, muggy day, the outside surface of the glass quickly dons a coat of water. This makes condensation look easy—just cool the water vapor and it will condense. But in fact the glass tempts water vapor out of the air not just by its temperature: the solidity of the glass helps trigger condensation. Water molecules must become more orderly to change from vapor to liquid, and the wall of glass is a helpful framework for the first venturesome vapor molecules that condense. There are no cold glasses of lemonade in the atmosphere. If the water were full of impurities, that would also encourage condensation, but the water vapor in the atmosphere is very pure water. In fact, vapor pressure could theoretically reach eight times saturation in the atmosphere, practically ensuring clouds would never form, if it weren't for a special ingredient of storms.

This ingredient is a scattered collection of microscopic particles in the atmosphere first identified by a Scottish scientist named John Aitken in the

1870s and '80s. Aitken developed a special chamber for counting particles in the air. He found that some particles attract water so well that they encourage condensation even when the air is not saturated. A newly condensed droplet grows faster than a pure-water droplet, because the particle dissolved in it continues to attract condensation. Best of all, Aitken found that the particles are practically everpresent in the lower atmosphere, as persistent as dust in your house, because, in fact, the impurities often are simply dust.

Another Scot, C.T.R. Wilson, followed Aitken's footsteps and helped establish cloud research. Born in 1869, Wilson was a sheep farmer's son who excelled in school, eventually becoming a medical doctor. One summer, while volunteering at the famous mountain weather observatory at Ben Nevis in Scotland, the 20-year-old Wilson was smitten with the sea of clouds below him. At Cambridge, he began researching cloud composition, using a dust counter adapted from Aitken's design. As moist air entered his lab through the dust counter mounted at the windward wall, he weighed the droplets collecting on a balance.

Wilson then tried to eliminate the dust completely during saturation, building the first cloud chamber. He was a skilled glassblower and, being notoriously cheap, he made everything himself. The cloud chamber was a glass cylinder with a tightly fitted glass piston. Wilson initially used no metal because he was studying atmospheric electricity and wanted to ensure that no unnecessary ions entered the chamber. With dust eliminated, the water vapor in Wilson's purified chamber didn't condense until it reached extremely high levels of supersaturation. The cloud chamber eventually helped Wilson win the Nobel Prize in Physics in 1927, honoring discoveries he made beginning in 1896. Wilson saw that X-rays made droplets in the cloud chamber even when all the dust was evacuated. Eventually he showed that the X-ray radiation ripped apart atoms and produced ions. These free ions became condensation nuclei for droplets in the chamber, as they do in the atmosphere.

These experiments proved the value of dust in storms. Thus the story of the ice storms of 1994, ironically, begins as all precipitating storms—in the hottest, driest places on Earth. In the Sahara desert, a typical grain of sand is very small. Line up a couple dozen of them end to end and they would make a row not longer than a mid-size ant. As small as they are, grains of sand don't get very far in the wind. They fall out of most near-surface airflows. But where terrain produces more violent turbulence—say, at the ridge of a sandy dune—gusts of wind can pick up the tiniest particles of mineral dust. These common quartz, feldspar, or mica particles (some as small as 1/100,000th of an inch in diameter) fall so slowly in the air that a stiff whirl of wind can carry them higher than a camel's head. At this height, the small eddies of air won't let them go, and they begin a long journey into the atmosphere.

Each year, the wind distributes about 100 million tons of desert dust across the globe. From the Sahara, the trade winds over the tropical Atlantic carry dust westward, toward the Americas. Sometimes, satellite sensors pick up the trail of this dust and follow it all the way across the Atlantic as a thick, dry cloud. Nonetheless, the atmospheric concentration of dust is slight: fill the Goodyear blimp a dozen times with Caribbean air, downwind from the Sahara, and you will have collected less than an ounce of aerial dust.

But the air streaming toward our shores carries more than dust. From the southwest, the subtropical jet stream feeding moisture into the February 1994 ice storm pushed air all the way across the Pacific. The trade winds hauling tons of Sahara dust toward the Caribbean and ultimately northward to the East Coast also traversed thousands of miles of ocean. Along the way, these streams of air built waves on the sea surface. The waves broke, creating a foamy splash of salt water. The smallest air bubbles in the foam eventually dissolved back into the ocean; they burst if they were too big (bigger than about a 100th of an inch across, that is).

Using high-speed photography, scientists have documented that the explosions of these bubbles are vitally important to storms. The dome of the bubble breaks into hundreds of tiny fragmentary droplets almost too small to consider. Immediately afterward, however, the sea surface closes in on the remaining crater, and a geyser of salty water droplets shoots up off the surface. These droplets lift tiny amounts of salt about as high as a shark's fin—high enough to get the salt circulating into the atmosphere. It takes billions of these droplets just to add an ounce of salt to the atmosphere— an incredibly small-scale process, indeed. But the oceans are endless, and waves countless. The breaking waves add a billion tons of salt every year into the atmosphere.

The salt, dust, and other aerosols (such as sulfates from fires and pollution) are light enough to stay in the atmosphere nearly forever. But in fact, the average particle manages to stay aloft only about a week. The vapor dissolved from lakes and oceans also manages to stay aloft for only about a week. This is more than coincidental; it is symbiotic. The atmosphere cleans itself daily of impurities by rainfall, but without the impurities rain would practically never fall. Probably 100,000 tons of salts and sand were embroiled in the February 1994 ice storm at any given time as it moved from Texas to New England. A large portion of that material ended up on the ground, though road maintenance workers would end up spreading more than 100 times that much salt to keep traffic moving through the snow and ice storms that winter.

If the air in a cloud is rich with nuclei, condensation happens easily once air cools to a small fraction of a degree below saturation. A grapefruit-size sample of such a cloud has three times as many droplets (about one million) as the city of Miami has residents. Other clouds have only 1/10th as many, especially if the droplets are larger (and the updrafts are willing). Either way, these are tiny droplets indeed. You'd have to stack about dozen of them to equal the thickness of this page. A thousand cubic feet of a typical cloud might have only an ounce of condensed water.

Droplets at this stage of cloud growth are far too small to fall out of the cloud. Initially, a cloud droplet might fall only half an inch a second in still air. Even a storm's weakest updrafts (a few inches per second in the warm front clouds) can carry the new droplets upward. By the time the droplets have risen a few thousand feet above the base of the cloud, they have grown considerably simply by accruing more and more vapor. As they climb higher, they grow to about four times their original size. As the droplets condense more vapor, however, vapor gets scarce. The droplets continue growing by merging with smaller, faster-rising droplets until they are about two-thirds the thickness of this page. Thus the droplets weed their smaller cousins out, getting bigger but fewer.

If the droplets can grow to about 1/250th of an inch, or a bit thicker than this page, then they will fall much, much faster than before. With a fall speed of more than 10 feet per second in still air, they can descend in most cloud updrafts. If they somehow form near the ground, or fall in the saturated air of a dense fog, light precipitation will occur. But such droplets are more likely to evaporate after venturing a few feet below a well-defined cloud base. Even a droplet four times thicker won't last more than a few hundred feet in dry air—not enough to make rain in most cases. In fact, a good rain from a high-based cloud most likely starts with droplets 10 times bigger than the minimum—1/25th of an inch in diameter.

For the first few decades of this century, meteorologists couldn't explain such phenomenal droplet growth to raindrop size. Condensation wasn't the answer. Clouds simply ran out of spare vapor long before droplets exceeded a 10th the diameter of a raindrop. And droplet merger seemed to take far too long to explain the heavy rains that fell from clouds only 30 minutes old. Besides, for droplets to merge, they had to collide. They only collided if larger droplets, which ascended slowly, were mixed with small droplets, which ascended quickly. Most clouds seemed to have fairly uniformly sized droplets. By World War II, however, a new theory of rain emerged that overcame these difficulties.

In 1911, Alfred Wegener published a textbook on meteorology in which he proposed that ice in clouds could help large drops form. Wegener was one of the most famous meteorologists ever to live, but not for his considerable accomplishments in the field. The German scientist also discovered continental drift. When Wegener first proposed it in 1912, colleagues mocked him. It took nearly half a century for the idea geologists at first called "utter, damned rot" to become the basis of modern plate tectonics.

Wegener died long before he won his stature among geologists. In 1930, the 50-year-old scientist led an expedition to Greenland. He heroically skied halfway across the glaciated island in the teeth of arctic winds to deliver critical supplies to his colleagues, who were stranded atop the 10,000-foot-thick ice cap. Satisfied that his men would survive the coming winter, he turned right around and headed toward the coast, but tragically succumbed to the elements halfway back.

Wegener proposed that rain was melted snow from high in the clouds. Others had guessed as much, but Wegener figured out how snow is an essential step toward most rainfall. Wegener noted a tiny difference in the saturation vapor pressure over water and over ice at the same temperature. He suggested that water would deposit itself on ice crystals at the expense of liquid drops if the two forms coexisted in the air. That water and ice might coexist at the same temperature may seem strange, but in fact water in free air is just as reluctant to freeze as vapor is to condense. In laboratories, in fact, scientists have been able to "supercool" water below −30 degrees Fahrenheit without freezing it. Atmospheric water is often supercooled, partly because the nuclei that work for freezing are different substances from those that work for condensation. Many of the best ice nuclei don't work well until water reaches 10 degrees or lower. (These nuclei tend to be minerals that are among the larger dust particles in the atmosphere.) Puddles on the ground freeze at close to 32 degrees because the water is dirty and the banks of the puddle offer a good starting point for crystallization. The purity of water vapor retards the freezing process in the atmosphere. (The puddle also has another advantage: its surface is much flatter than that of tiny droplets, meaning there is far less surface tension to impede the freezing process.)

Tor Bergeron, the discoverer of the occluded storms, made another great contribution to meteorology by establishing the truth of Wegener's proposal about ice and supercooled water. Bergeron's wide-ranging impact on weather science was born of a deep passion for the vagaries of the atmosphere. At the age of 17 Bergeron checked the barometer every day for a month for a school assignment. Observing the weather soon became his obsession. Bergeron developed theories about visibility in the atmosphere from his early observations. A few years later he related visibility to wind shifts while working as a meteorological assistant for Johannes Sandstrom, the Bjerknes pupil who taught Bergeron streamline analysis.

We have already seen how crucial Bergeron was to the Bergen group. In 1922, between stints in Bergen, Bergeron vacationed at a health spa on the hills near Oslo. Here he took frequent walks on the roads that wind through the local fir forests. Stratus clouds frequently blanket these slopes at about 1,500 feet elevation, wrapping the trees in a cool, exhilarating fog. On some days, Bergeron noticed he was walking through the dense cloud. On other days, when the temperature was cooler (at least 10 degrees below freezing), the fog lifted overhead, though not far above the treetops.

Bergeron looked up and saw that when the fog lifted, snow covered the treetops that were once enshrouded with cloud. Just as Wegener had said, the air was more easily saturated next to the snow than next to the water droplets of the fog. He realized that the fog droplets evaporated, which saturated the air with respect to snow. That same moist air then gave back some of its vapor, freezing onto the snow. This lowered vapor pressure enough to encourage more drops to evaporate, continuing the cycle. Soon, the cloud around the trees cleared, and the snow thickened slightly. If instead no initial layer of snow coated the trees, the fog remained plentiful and descended to the road.

It took six years before Bergeron did anything with this observation, but when he did, in his Ph.D. thesis of 1928, he developed one of the most significant meteorological theories of the century. Once a few ice crystals form in a cloud, the water droplets begin to shrink. Ice crystals in a cloud gather up water vapor just as fast as the water droplets can get rid of it. Theoretically, a single crystal can be a janitor for a sizable region of cloud, mopping up all the water droplets in a matter of minutes. Bergeron showed that under the right conditions, one small ice crystal accrues the water of 1,000 droplets in about 20 minutes.

The onset of crystallization in a cloud is sometimes visible from the ground. As a small puffy cumulus cloud grows upward, it initially looks like a large head of cauliflower. But when the cloud reaches a significant altitude, where air is very cold, it rather quickly becomes a gauzy mush, like cotton candy. The cloud has reached an altitude where it is cold enough for ice crystals to form, and the crystals blur the shape of the cloud.

Above the mid-Atlantic states on February 9, 1994, crystals probably formed on dust from the Sahara and other sources in the air that rose at the nearly stationary warm front. The crystals grew into large snowflakes, first by the deposition of vapor onto the crystals. Later, as the flakes expanded beyond about a 50th of an inch wide, they started sticking together. In this particularly moist maritime air, the flakes probably stuck together quite readily. (In fact, similar moist air in 1887 produced snowflakes that were an incredible 15 inches in diameter in Montana, a record achieved by an amalgamation of probably well over 100 individual crystals.) Within about 15 minutes, the snowflakes grew large enough to fall out of the clouds at speeds of a few feet per second.

As they fell, the snowflakes far north of the front encountered cold air spun off the high pressure sitting off the New England coast. They floated down to the ground, evaporating and shrinking as they fell, but remained large enough to accumulate a foot of snow as they piled on lawns and streets. On the 9th and 10th, the snow fell heavily in part because of the circulation of the storm. The air swinging clockwise around the high pressure was cold, but it was warmed from below by the ocean. As it approached shore, the warmed bottom margin of air fell far short of saturation and picked up plentiful ocean moisture. Buoyed by the heat, it rose and formed low clouds that then cooled quickly over the icy New England landscape. This change of environment began to wring the ocean moisture out of the clouds—an unusual case of "ocean effect" snow. Two more feet fell on coastal towns.

A more common version of this snowfall, lake-effect snow, had been remarkable the previous month. January's storms kept pumping waves of cold Canadian air southward across the Great Lakes. Each blast of northerly and northwesterly air passed over the as-yet-unfrozen waters of the lakes and guzzled the rich supplies of moisture. The lake-induced clouds then dumped snow relentlessly on southern shores, even though storm centers were far, far away. Houghton, Michigan, got six feet of snow during the month. Two years later, during the prolific winter of 1995–96, Marquette, Michigan, got 250 inches of mostly lake-effect snow.

Lake-effect snow, if exacerbated by a steeply sloping shoreline, can cause even more astounding totals. Usually communities along the shores of Lake Erie and Lake Ontario gather well over 100 inches of snow before ice atop the lake mostly shuts down the moisture machine. Syracuse, New York, often leads the nation's major cities with snowfall, and in 1992–93 had 192 inches of snow. In Buffalo, lake-effect snows are so frequent forecasters have begun naming them the way hurricane forecasters name tropical storms. Nearby, at Montague, New York, on January 11–12, 1997, an incredible 77 inches of snow fell in 24 hours. The crystals in lake-effect snow, falling from very cold air, tend to be small and to settle fast, however. What could have been a national one-day snow record fell into dispute. By the end of the day, the snowpack had grown "only" 51 inches. Nonetheless, it's easy to see how a community like Hooker, New York, on the lake shore, reported more than 300 inches of snow in 1993–94, still far short of its record 467 inches.

That record was set in 1976–77, the winter Buffalo established itself as the epicenter of lake-effect snow. The city had just shy of 200 inches that year, but it was the effect that was so chilling. Most of the snow fell during a 53-day streak that winter, settling quietly, without much harm. Then came a blizzard on the afternoon of January 28. Winds gusted to 85 m.p.h., sweeping a foot of new snow across the snowpack, whipping up 25-foot drifts. People were buried in their cars trying to beat the storm home during rush hour. The lucky ones were stranded at their offices for several days. Twenty-nine others died. For the first time in the nation's history, a city was declared a federal disaster area due to snow.

Lake-effect snow is impressive but localized. Some adventurous meteorologists try "chasing" lake-effect storms the way their colleagues chase tornadoes. The snow-laden clouds form over the lake, usually in bands perpendicular to shore. The bands mark the heated, rising currents of wet air, which are rolled into long tubes by prevailing onshore currents. A snow chaser's goal is to guess where the first flakes will fall, based on where the cloud bands form. In between the bands, hardly a flake might fall. In one legendary lake-effect storm in Chicago, in 1903, forecasters sat in their office admiring the sunshine when angry residents called from the suburbs to complain about snow. The meteorologists had no idea that 14 inches was falling just a few miles away.

In the February 1994 storm, where you stood also made a difference in what weather you got. The snowflakes far south of the frontal zone did not fare nearly as well as the ocean-effect crystals on the Massachusetts shore.

Over the Carolinas and southward, flakes began to melt fast as they hit mild 40- and 50-degree air. They became large, cold raindrops and their fall suddenly accelerated from less than 10 m.p.h. to about 60 m.p.h. when they liquefied.

This transformation of precipitation aloft was one of the early discoveries of the radar age in meteorology. Falling snow doesn't reflect the beam of a radar very intensely, so it makes a relatively weak blip on the display screen. But as the flakes melt and acquire a covering of water, they yield a much more intense radar return. A bright band appears on the radar screen, signifying the melting level in the atmosphere. Below it, the return is again weaker, because the raindrops accelerate so much. The bright band helps confirm Bergeron's rain theory (now called the Bergeron–Findeisen process, also honoring the German meteorologist Walter Findeisen who during the 1930s and '40s did much work on freezing and precipitation in Prague, but who disappeared near the end of World War II).

The snowflakes along the front paralleling the Mason–Dixon line in the February storm also melted to rain, but near the ground they encountered cold air beneath the rising warm air. In other words, there was a sandwich of warm air between the high-altitude cold and the cold near the surface. Where the surface cold air was thick enough, the rain refroze and hit the ground as sleet, making piles of tiny white ice balls. Where the cold air was not very thick, the rain supercooled, becoming what is called freezing rain. When it hit the ground (or a tree or a power line—anything solid), the supercooled drops froze immediately, turning the Southeast into a vast skating rink.

The storm produced a tremendous amount of freezing rain because the cold air mass squatting northeast of it was relatively shallow for the season. East Coast storms thrive on the temperature differences between warm Gulf Stream waters and the cold air streaming off the North American continent. The worst East Coast snow storms often pair up with a high-pressure mass of air over the Northeast. As a low-pressure center of a storm moves up the coast toward this high, the pressure contrast intensifies in a narrow zone; the result is usually northeasterly or easterly winds blowing intensely toward the shore in between the high and low centers. The cold winds whip up big waves, spelling trouble for coastal communities. In February 1994, the usual high over the Northeast was centered offshore rather than over land. Temperatures in upstate New York hovered around zero, but the cold didn't work its way southward decisively. That's why the cold air was relatively shallow at the warm front and, as a result, much of the precipitation at the front was freezing rain rather than snow or sleet.

The sensitivity of precipitation to the depth of cold air beneath the warm front makes freezing rain a tricky phenomenon to forecast. But since cold air eventually retreats before the warm front, freezing rain is usually a brief intermission between rain and snow. In January 1998, however, freezing rain pummeled Quebec and northern New England for nearly five straight days. In this case jet stream mechanics were paramount, because a persisting convergence of jets over eastern Canada maintained high surface pressure. Around the high, bitterly cold air swept southwestward through the St. Lawrence seaway and into Montreal and Ottawa to the north, and Vermont and northern New York state to the south. Cold air persistently advanced against all odds under a few thousand feet of warm air that had flown in all the way from the Atlantic off the Carolinas. While Canadians were getting pelted with ice, the warm air opened cherry blossoms prematurely in the nation's Capital and set a record with a 70-degree high in Newark, New Jersey. The warm air clashed into widespread storminess by January 9 as up to 13 inches of rain fell in the South, causing floods that killed two people in Alabama.

Further north the result was devastating. Officially, four inches of precipitation turned to ice as it pelted the area. Nearly a million people were without power in mostly rural areas from Watertown, New York, to upstate Maine. People sought shelter at hospitals, schools, shopping malls, even a prison, but many others were stranded. Even emergency workers couldn't negotiate the icy, tree-blocked roads. High winds after the ice merely added to the debris. A New York state trooper compared the scene in Watertown to the aftermath of a B-52 bombing run. Even the local hospital lost power for seven breathtaking minutes. Nurses pumped air for patients on respirators until a backup generator started up.

The storm was one of the costliest natural disasters in Canadian history, and the army mobilized 11,000 soldiers to help out, an unprecedented peacetime deployment. Another three million people—nearly half the region's population—were without power. Many city dwellers went without power for a week; farmers and others in rural areas endured the dark far longer. People opened their homes and ran their wood stoves for strangers whose own houses were getting too cold. With fire crews unable to negotiate the tree-blocked, icy streets, residents put out their neighbors' blazes when generators shorted and fireplaces spewed errant sparks. Many people risked carbon monoxide poisoning trying to get warm by idling their cars in the garage as temperatures after the deluge dipped below zero before power could be restored. To fix the downed power lines, Hydro-Quebec had to buy more than 30,000 new poles.

After the storm it was treacherous to walk the sidewalks of Montreal, not just because of the slick surface underfoot, but also because deadly chunks of ice shook loose from branches and ledges above, without warning. More than three-quarters of the trees in famous Mount Royal park suffered damage. The following spring, tree pruners sadly had to remove 10 percent of the trees on public property throughout Montreal. Property insurers in Canada opened their pockets to the tune of nearly $1 billion, and the utilities spent about the same on the recovery.

Fortunately, not all storms ice the landscape. But Tor Bergeron's explanation that all rain starts as ice in the sky was a relief for many meteorologists. It certainly made it possible to describe the evolution of precipitation in a storm like the February 1994 ice, sleet, and snow extravaganza. The theory ran into some immediate difficulties, however, and cloud physicists are still sorting out the details.

Early on, scientists homed in on one of the weaknesses of the Bergeron–Findeisen process, which was the general uniformity of droplet size and distribution in the cloud that Bergeron had assumed. Scientists discovered that as soon as one ice crystal formed, the uniformity began to deteriorate. Even worse, however, scientists found hard evidence when they flew through clouds to collect crystals. Concentrations of crystals varied anywhere from a few per gallon of air to several hundred per gallon—all within the same cloud. The ice crystals sometimes were 10 or even 1,000 times as numerous as the nuclei supposedly necessary to form them, especially in older clouds.

One answer to these worrisome findings emerged by the early 1970s—that not all ice crystals form around traditional nuclei. New ice crystals might form when fragments of crystals break off during midair collisions. This process seems especially important in some clouds over the High Plains, where crystals in the very cold air grow many branches, like a classic snowflake. Another theory holds that as a droplet freezes, it forms an outer shell of ice first. As the inner water freezes, it expands, shattering the old shell, fragments of which begin new crystals. Another theory came from lab experiments with riming. Rime is the ice formed when supercooled water freezes on impact. When ice crystals grow fast by riming in the clouds, the freezing of the outer coat of water warms the interior slightly. Water underneath squeezes out through cracks in the rimed outer coat, then quickly freezes, forming microscopic icicles. The delicate needles break off easily during further riming and ice collisions. Nearly every incoming supercooled droplet has the potential to break off 10 spikes from the host crystal, easily explaining the exponential multiplication of crystals in the cloud.

Investigating the role of ice crystals in rain formation led scientists to many remarkable findings about clouds in the past few decades. For instance, some largely warm clouds (warmer than freezing) still make rain. Many rainy sheets of cloud along warm fronts seed themselves with abundant crystals. The relatively flat and unremarkable nimbostratus may be quite warm on the bottom but still be colder than 32 degrees Fahrenheit near their tops. The ice forms only near the top of the cloud, but because air rises extremely slowly in nimbostratus (usually around 1 m.p.h. or less), the crystals fall through the cloud. For a short distance, perhaps a few thousand feet, vapor deposits itself as ice on the crystal. A nimbostratus forms rain in a leisurely fashion—sometimes it takes three hours from cloud formation to precipitation on the ground.

Sometimes one cloud seeds another. Warm-frontal air rises gradually, making a stratus cloud. But small cells of fast-rising air sometimes lurk within these thick layers of cloud. These cells form concentrated frontal rain bands often about 30 miles wide, parallel to the front. Updrafts in the embedded convective towers are strong enough to form large crystals that eventually fall out and settle like confetti into the warmer layer of cloud below. Here gentle stratus updrafts provide plenty of water to deposit onto the ice crystals from above. The growing crystals eventually melt to make locally intense rain. The vast majority of the water in the raindrops comes from the lower stratus deck, but the crystals from above are essential.

You may have noticed as you watch a thunderstorm pass by that the rain under it is highly variable. When the rain shaft catches you, sheets of intense rain thick enough to drown a cat seem to blow in the wind, while in between are respites of merely heavy rain. Scientists flying into thunderstorms have noticed that large droplets inhabit the very top of the strong updrafts in the growing cloud turrets. This plentiful water makes ice crystals grow extremely fast as the turret reaches its peak height. The turret may boast hundreds of crystals per gallon in pockets of intense ice growth only about as big as a three-bedroom house. As the ice grows very heavy, it falls in strands of large crystals that grow even larger as they meet supercooled water rushing upward. In the Arizona desert, or anywhere else where you are lucky enough to see such storms from a distance, the backlit rainshaft below the cloud base will appeared striated because of these dark, extra-intense rain strands hanging from old, fast climbing cloud turrets.

While the distribution of ice crystals in the cloud can shape the rainfall below, some meteorologists couldn't accept the Bergeron–Findeisen process as the only mechanism for rain. The theory was far better than any other when Bergeron announced it in the 1930s, but that didn't convince everyone that warm clouds couldn't also produce rain. Meteorologists finally obtained evidence of it after World War II. Using radar to study thunderstorms in the 1940s, meteorologists discovered specific examples of summer storm clouds that form ample precipitation long before climbing into the freezing reaches of the atmosphere. Clearly, tropical rainfall must form similarly. Now scientists had to figure out how warm rain forms.

Warm rain requires a lot of water in a cloud and a very uneven mixture of drop sizes. An updraft is also necessary—one strong enough to detain a droplet while it grows early on, then not too strong to prevent the droplet from falling and bumping into other droplets. As large droplets begin to emerge, the growth by collision is more and more efficient. Laboratory studies show that when droplets collide, they can't stick together unless they are very different sizes. Similar droplets brush by one another in relatively slow motion, whereas a very large drop bears down speedily on a small one. Some of the small droplets ram straight into the big drops. Others are carried around the drop in the sweeping airflow, but get sucked into the back by turbulence above the big drop. In the 1940s, the Nobel Prize winning chemist Irving Langmuir demonstrated these processes with simple mathematics. One of the ways some droplets get much bigger than their cousins is by using big nuclei. Sea salt turns out to be perfect for warm rain, hundreds of inches of which fall each year in parts of Hawaii. Sea salts don't work just in the tropics or over the ocean—they exist in maritime tropical air that feeds storms over California, the Midwest, Florida, and the East Coast.

Drops have to be big just to make it to the ground. A rain cloud can last an hour, but droplets evaporate in half that time, especially because their parent clouds tend to turn themselves inside-out, exposing growing droplets to dry, outside air. As a result, warm rain droplets must get a head start on growth while rising in the relatively favorable, moist environment of the inner cloud. At 1/100th of an inch in diameter, the drop can withstand a long fall through dry air, if necessary, when it tumbles out of its parent updraft into the waiting arms of a younger updraft in the cloud. If a warm cloud is too short, or the updrafts are too fast, the young droplets might not grow enough early on to make rain. But add a few hundred feet in height to the updrafts, and you have a downpour.

A young researcher in England named Frank Ludlam helped establish these facts of warm rain by making a simple mathematical model of a raindrop falling in an updraft. He showed that the size of the droplet emerging at the top of an updraft basically determines the size of the raindrop hitting the ground later. This means the height of the cloud and the updraft speed are critical conditions of warm rain. With an updraft of a few feet per second, a cloud needs only to be about 6,000 feet tall to make warm rain. A mild thunderstorm updraft of about 25 feet per second means the cloud has to be three times as tall to make warm rain. Ludlam also was able to use mathematical models to calculate how thick a cloud would have to be, given an updraft speed and a base temperature, to

Vincent Schaefer creates snow crystals in a refrigerated box in 1947 by supercooling his own breath and then seeding the vapor with dry ice. Later, his General Electric research colleague Bernard Vonnegut found that silver iodide seeded hexagonal ice crystals even more effectively. Photo courtesy of General Electric Research and Development.

generate rain with the Bergeron–Findeisen process. Subsequent research using more sophisticated models followed Ludlam's basic design.

Despite the discovery of warm rain, the Bergeron–Findeisen process marked a new maturity in cloud physics, and it inspired a whole new field of meteorology after World War II—rainmaking. Scientists got deeply involved in seeding the clouds after the war because of the work of a group of researchers led by Langmuir at General Electric. Langmuir and his assistant, a machinist named Vincent Schaefer, were principally engaged by the military to improve gas mask filters, but they also investigated the effect of supercooled water on radio static and then on airplanes. The best place to study supercooled water in the clouds is in the clouds themselves. The leading observatory "among the clouds" in the United States is atop Mount Washington in New Hampshire. Mount Washington, at an elevation of 6,288 feet, is the highest point east of the Rockies. It is often shrouded in howling winds and blinding snows and clouds for days at a time. Winds there during a storm in 1932 set a record of 232 m.p.h.—not just a gust but a sustained speed recorded with a standard anemometer for more than a minute. The meteorologists who spend long days in their laboratory encased in ice at the summit must venture out to collect their rain bucket occasionally for measurements. When the walk is slicked with ice and the wind tops 100 m.p.h., which is frequent, newcomers often get a taste of mountain humor: the game is to make it to the bucket and bring it back safely without slipping. Very few people stay on their feet for more than a few seconds.

Between 1943 and 1945, Langmuir and Schaefer spent a lot of time on Mt. Washington and invented airplane wing surfaces that would slow riming even at temperatures below zero Fahrenheit. One riming defense is to crystallize the supercooled water before it fuses to the wing. The scientists tried dirt, quartz, pollen, coal dust, and other potions to get supercooled water to crystallize. One day in July of 1946, Schaefer was working in the G.E. lab but found it was too hot to get his cold chamber (an open freezer) to cool down for his experiments. Getting impatient, Schaefer found the handiest cold substance he could find: he dumped dry ice (the solid form of carbon dioxide often used to make stage smoke) into the cold chamber. He got a surprise: the supercooled water in the chamber instantly turned into fairy dust—a bluish cloud of ice crystals. Excited, Schaefer repeated the trick over and over. It was simply serendipitous, yet he had found the magic ingredient needed to turn water into ice. "By scratching the dry ice with a nail so that tiny fragments of dry ice would

The first successful cloud-seeding flight by General Electric scientists digs a race-track trough where fast-growing crystals sink out of the clouds. Photo courtesy of General Electric Research and Development.

fall into the cloud, I could produce a spectacular stream of ice crystals under complete control," he explained.

It didn't take long for Schaefer to realize what his discovery was worth. The dry ice could kick-start the entire Bergeron–Findeisen process—the growth of raindrops in a cloud. On November 13, 1946, G.E. scientists flew up above a stratus cloud filled with supercooled water droplets and dumped dry ice on it. Just as in the cold chamber, the seeded area quickly turned to ice crystals, which fell and left a sunken track in the cloud.

Bernard Vonnegut, another soon-to-be-famous member of Langmuir's group, figured it was possible to find a substitute compound to seed clouds. The secret was in part to mimic the shape of ice crystals. Ever since ancient Chinese scholars called snow six-petaled blossoms, people have been curious about the shape of individual ice crystals. As it turns out, ice always forms in some basic hexagonal plan. Some crystals are hexagonal plates, others are starlike crystals with variations on six branches. Some are

An inventive and persistent Wilson Bentley illuminated the infinite hexagonal variety of snowflakes by capturing and photographing them under a microscope on his farm in Vermont.

so thick that they make six-sided columns, like pencils. A Vermont farmer named Wilson Bentley developed a technique for photographing these flakes before they melted. In his spare time, early in this century, Bentley photographed thousands of variations on the basic forms. In his catalog of snow images, of course, no two crystals were alike. Not until 1989 did ice researcher Nancy Knight of the National Center for Atmospheric Research, in Boulder, Colorado, finally isolate the first pair of identical snow crystals. Vonnegut experimented with chemicals that had hexagonal crystals close to the shape of ice crystals, and in November 1946 he found that silver iodide worked. This chemical is not natural to the atmosphere but is an effective nucleus for vapor deposition at about 10 degrees Fahrenheit and for freezing liquid water up to about 24 degrees.

Langmuir got a bit carried away with optimism for the new cloud-seeding methods. Already in 1947 he predicted altering thunderstorms, eliminating hail and freezing rain, saving farmers millions of dollars. With dry ice, he claimed, it would take only 15 minutes to clear a sizable gap in the dreary winter cloud cover over northern communities. Schaefer thought that rainmakers might turn a watery cumulus directly into an icy cirrus before the growing cloud had a chance to generate lightning and start forest fires. (Nearly three decades later NOAA scientists succeeded in reducing lightning intensity and frequency by seeding thunderstorms with aluminized Fiberglass chaff dropped from planes.)

In 1949 Weather Bureau chief Francis Reichelderfer sent one of his scientists, William Lewis, to watch Langmuir seed hundreds of clouds in New Mexico. Things had gotten a little dicey the previous fall. Project Cirrus, Langmuir's military-funded seeding project, started enhancing clouds with silver iodide on the same day flash floods struck the Albuquerque area. The project used an old B-17 bomber for seeding the area, but the scientists also rigged a silver iodide smoke generator atop their Oldsmobile Coupe and chased after clouds across the desert. In one encouraging test, snow fell for two hours near Santa Fe only about 25 minutes after scientists upwind had seeded the clouds with silver iodide smoke.

Problems mounted quickly for the cloud seeders. For starters, Lewis concluded that the silver iodide nuclei weren't necessary to make rain, and that natural nuclei were abundant already. He believed the rainmakers weren't adding anything the clouds couldn't make on their own. A few years later, the Weather Bureau investigated complaints from farmers and ranchers in the West that commercial cloud seeders were hired by people who wanted rain when others nearby didn't want it. This was a potentially widespread problem: in 1951, rainmaking covered more than a tenth of the area of the contiguous states. Heavy snows that year threatened to damage mine shaft openings in Colorado. The miners turned around and petitioned the governor to review cloud-seeding practices that might contribute to the snow.

At the next meeting of the American Meteorological Society rainmakers themselves were at loggerheads over the new technology. Langmuir presented "conclusive proof" that New Mexico seeding affected rainfall over an area extending eastward all the way to the Mississippi River and beyond. This was a fairly incredible statement, in part because silver iodide seemed ineffective after prolonged exposure to sunlight anyway. But Langmuir was struck by the timing of rain hundreds of miles to the east: it seemed to correspond exactly to the timetable of his seeding experiments in the West, with an appropriate lag time, of course. Other meteorologists believed the seeding worked but that experiments were

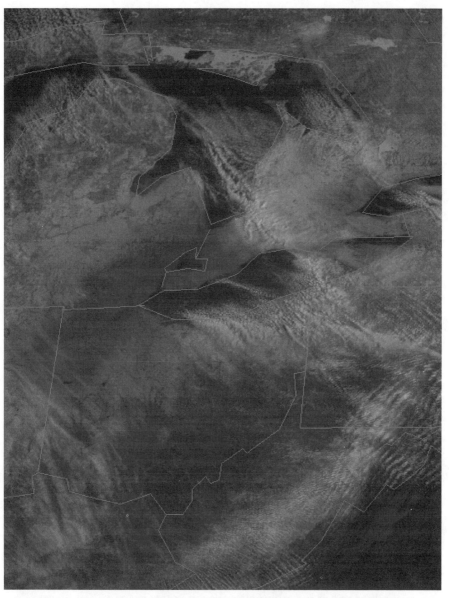

From space they are thin strokes of cloud brushed lightly southeastward from Lake Huron past Lake Erie, but on Earth these lake-effect bands pick up moisture from warm waters, then unmercifully dump it by the foot as snow on cold land. The narrow rows ensure that lake-effect storms will be local but long-lasting. NOAA satellite image courtesy of Lee Grenci.

nonetheless mostly inconclusive. In 1952, satisfied that silver iodide often helped make rain, the Department of Defense and General Electric ended Project Cirrus.

Rainmaking didn't stop there; it flourished, and it remains a viable business for some meteorologists and an active area of research in some countries. For a time the military worked with cloud seeding as a weapon. In 1966 the Air Force tested silver iodide and lead iodide flares over Laos. The next year they began a five-year project, called Operation Motorpool, designed to enhance monsoon rain clouds. Air Force pilots flew more than 2,600 cloud-seeding sorties, attempting to create enough rain to make the Ho Chi Minh Trail impassable. Some intelligence experts claim the project indeed increased rain in some areas by nearly a third. But in 1971 newspapers blew the cover on the secret project, and Senator Claiborne Pell claimed the Air Force had killed thousands of people with man-made floods. "If we permit weather modification, we've really opened up a new Pandora's box," he told reporters. Pell began a successful campaign for a treaty to ban environmental manipulation in war. Later, after the military briefed him privately on Motorpool, the senator supposedly changed his accusations somewhat, saying, "An elephant labored and a mouse came forth." In a more benign use of the technology, in 1969, the Air Force cloud seeders attempted to alleviate a drought in the Philippines. They also flew rainmaking missions two years later over Texas during a dry spell.

If the news leak damaged military prospects of developing rainmaking, it didn't stop private rainmakers domestically. By the early 1970s, rainmakers were working over 146,000 square miles—5 percent of the total area of the country, including Alaska. One cloud-seeding project over South Dakota in 1972, however, demonstrated the volatility of the cloud seeding, if not from a scientific point of view, at least from a political one.

Department of Interior rainmakers were seeding clouds near Rapid City, South Dakota, on the night of June 9, 1972. Nature was taking care of its own business not far away. Thunderstorms built up over a small watershed in the Black Hills around Canyon Lake, west of Rapid City. The storms lodged against the rugged terrain and wouldn't budge. It only took about six hours for the rain to accumulate into a highly localized deluge: tens of millions of drops pummeled every square foot of a 100-square-mile area. The 10 to 15 inches of water cascaded down the slopes to Canyon Lake. Such rainfalls are not unheard of: Holt, Missouri, got a record 12 inches of rain in 42 minutes in June 1942. But Rapid City lay vulnerable to this water, beneath a straining dam and a vital spillway clogged with debris.

At 7:30 that evening, the worried mayor of Rapid City was telling people to get out of the flood plain running through central Rapid City around Canyon Creek. It was dark, and few people saw the danger of the rising creek. At 10:45, however, the dam broke, and deluge became disaster. In the black of night, more than a billion tons of rainwater coursed through the small town. The water suddenly rose 12 feet and topped the banks of the creek higher than a man's head. Some people spent the whole night riding logs in the torrent. Others climbed trees but couldn't hold on and were swept away by the water. In the dark, children screamed for help from rooftops. Afterward the center of the town was a mush of mud with a few roof tops and car hoods peeking through. The death toll stood at 238, and more than 1,000 homes were destroyed.

When they found out about the cloud seeding that night, the people of Rapid City sued the government. Legal action took them nowhere: the rainmaking likely didn't affect the flood. But, they also smartly moved the center of town to higher ground and cultivated a greenway around the deadly river. Now prepared for the next freak of nature, Rapid City learned a lesson about water and weather that has been duplicated each year around the world.

CHAPTER SEVEN

Terrible Ups and Downs

When Lieutenant Colonel William Rankin awoke on Sunday, July 26, 1959, the sky was bright and calm over South Weymouth, Massachusetts. The Marine Corps fighter pilot had no idea what was in store for him that day. That afternoon meteorologists told him thunderstorms topped 30,000 feet over Norfolk, Virginia, but otherwise his flight from Boston to South Carolina should be smooth. It was far from smooth. But then who would have expected a face-to-face encounter with a merciless thunderstorm? Rankin would see what few people ever see and almost no one has survived to describe.

Rankin wisely rode the afterburners of his F8U Crusader to 44,000 feet to cruise over the rough weather. Approaching Norfolk at close to 6 p.m., he could see darkening cloud towers glowing with lightning. The storm had grown since he had taken off, so he climbed above 47,000 for extra clearance. That's when the plane gave Rankin a jolt as it started to rattle. A warning light flashed, filling Rankin with dread: the engine had seized up. Now without power, the plane was about to plummet out of control. Rankin had no choice but to eject into the −70-degree Fahrenheit unknown above the storm.

The shock of the cold stung him as the thin air tugged at his eyes, burst his ears, and bloated his stomach with excruciating force. In about 15 seconds Rankin tumbled spread-eagle into the cloud, falling at about 100 m.p.h. He couldn't tell how fast he was falling, but he knew at least it would be over soon. Rankin's chute would open at 10,000 feet. He began to breathe easier through his mask as the air warmed. "I felt as if I were suspended in a soft, milk-white substance, a huge amorphous easy chair."

Then the chute opened—too soon. Rankin was still inside the cloud, and with the parachute the storm took Rankin for the ride of his life. Baseball-size hail pelted his body, his life probably saved by his helmet. The thunder shook him violently and the blue sheets of lightning blinded him temporarily. Torrents of rain made it nearly impossible to breathe. Still aloft 20 minutes after bailing out, Rankin realized he might never escape the storm. Over and over the updrafts shot him upward with a jarring blast, then dropped him with a gut-wrenching lurch into downdrafts. In some of the updrafts Rankin felt like a shell fired from a cannon. The pure white around him transformed into "an angry ocean of boiling clouds—blacks and grays and whites ... digesting one another."

Finally the storm spit Rankin out and he crashed to the ground at more than 35 m.p.h. He had been in the air for 40 minutes. Shivering, bruised, and bloody, he had miraculously survived. He lived to tell what makes a thunderstorm the nastiest engine of weather on the planet. The updrafts and downdrafts that swung Rankin about like a yo-yo drive deadly, severe weather. They lurk behind practically everything dangerous in volatile storms—hail, lightning, heavy rains, and high winds. They make the thunderstorm a special breed of cloud—one that can ignite a field, flood a canyon, bring down an airliner, or flatten a forest, all in a matter of minutes.

In a storm all the atmosphere's pistons are firing in a well-timed vertical frenzy. Yet it all might start over a lonely patch of warm grass or dirt. A rising column of spiraling air perhaps only 100 feet across marks a patch of ground hotter than its surroundings. Local winds collect here along the ground. You might see a solitary bird of prey. Its wings motionless, the hunter circles tightly within the updraft to stay aloft. This is a telltale sign of an updraft, perhaps the beginning of a thunderstorm, yet the heated roots of updrafts were mostly a mystery until this century. Orville and Wilbur Wright were among the few to guess that circling birds take advantage of warm rising air (also called a thermal). They figured a glider pilot could also stay aloft for hours by steering from thermal to thermal. The idea was nearly forgotten until a quarter-century later, when a scientist in North Africa thought to test the idea with a thermometer: he ran under circling vultures and found that in spots underneath them the temperatures were a degree or two higher than the surroundings.

Hot spots are only one way to kick-start an updraft. Over the vast thermal uniformity of the tropical ocean, millions of updrafts develop vigorously every day without hot spots. Clearly, other atmospheric mechanisms also make the surface convergence that triggers an updraft. These include cold fronts and, in the tropics, the trade winds from the hemispheres which meet regularly not far from the Equator. Large-scale sea breezes converge over Florida nearly every day, making the peninsula the thunderstorm capital of the country. Some towns in the Sunshine State get thunderstorms 100 days a year.

In India, the monsoon similarly converges ocean and land air over the subcontinent, firing up colossal storms and rains. Some severe storms go one better than mere convergence. Air streaming northward from the Gulf of Mexico, for instance, will often pool for days over the continent because it is capped by subsiding air from upper-level high pressure centered off Bermuda. Few clouds form and the air bakes intensely day after day. When a cold front finally plows through this Gulf air, or a rare, intense hot spot finally blows through the cap, the updraft soars with stored energy like a geyser at Yellowstone.

At several thousand feet or more, the vapor begins to condense in an updraft. A cumulus is born. A number of thermals banding together release heat and fuel the growing cloud. Most cumulus don't last long, but their updrafts are significant nonetheless. In the 1920s, many scientists still couldn't believe that cloud updrafts exceeded a couple miles per hour simply owing to surface heating. Then in 1926 a glider designed to descend at a rate of 4 m.p.h. in still air got blown several thousand feet up into a thunderstorm in a matter of minutes. This meant the updrafts were 10 or 15 m.p.h., maybe swifter. Intrigued, German glider pilots awakened to the possibilities of thermals and rediscovered the Wright brothers' idea of soaring from cloud to cloud. One man flew his glider nearly 90 miles that way in 1929. But pilots were often frustrated by the technique. The clouds often evaporated in the few minutes it took to fly over to them.

Not all clouds evaporate right away, of course. A rising parcel of moist air can easily become warmer by 10 degrees Fahrenheit than its high-altitude environment simply owing to heat released by condensation. This latent heat release is energy for wind and updrafts: a single cumulus releases latent heat energy equal to the blast of a 1,000 pounds of TNT every minute. Tropical clouds transfer so much solar energy to the climate system in this way that they drive the atmosphere everywhere. In fact, tropical cumulus clouds are overkill: they release 40 or 50 times as much energy as needed to drive the world's winds. With such copious amounts of heat, it would seem the sky's the limit for these buoyant clouds. At least the stratosphere should be the limit, and it often is for a 40,000- to 60,000-foot storm cloud. At that height, air in the updraft finally stalls in a warmer environment. But not every cloud is so efficient, fortunately, or we'd have many more thunderstorms.

In the tropics, actually, very few of the millions of building cumulus climb past 7,000 feet. This altitude has a nearly permanent inversion—a buoyancy-killing condition with warm air over cooler air (or over air that would be cooler if it rose). World War II exposed many meteorologists to tropical questions, and one of the biggest was why relatively few tropical clouds break through the inversion. There seemed to be plenty of heat below them. Mathematically, it made no sense. For 100 years meteorologists had treated the air in updrafts as isolated blobs called air "parcels." It was relatively simple to compute how high a parcel of air would rise based on its temperature, its humidity, and the characteristics of its environment. But the parcel model didn't seem to work for these tropical cumulus—it grossly overestimated the actual ascents observed. A scientist from MIT's Woods Hole Oceanographic Institute, Henry Stommel, figured out a way to improve the mathematical calculation and in the process brought a new level of sophistication to cloud study. Stommel compared rising cloudy air not to a parcel but to a high-speed stream of flow—a jet—disrupting a quiet body of fluid. The jet forges ahead into the rest of the fluid, but at the same time creates turbulence at its edges. The turbulence mixes surrounding fluid into the jet, diluting and slowing the flow. If an updraft is an upward jet of air, then it dilutes itself, weakening the cloud's buoyancy. The dilution by dragging in surrounding air is called entrainment. By 1950 Horace Byers (who succeeded Rossby as head of the meteorology department at the University of Chicago) already believed that Stommel's model of an entraining updraft had "a profound effect in modifying all our thinking about the thermodynamics of the thunderstorm."

At Imperial College in London in the 1950s, Frank Ludlam and Richard Scorer came up with a slightly different analogy for entrainment. They said clouds represent a series of bubbles of air rising in the atmosphere. Scorer demonstrated how this works by trapping two-inch bubbles of colored fluid with an inverted can in a four-foot-deep tank of water, then releasing the bubbles and watching them rise like an updraft. Mixing occurs at the top and around the sides of the bubble; then the mixed outer layer slips around toward the bottom of the bubble. The slippage of water turning over toward the bottom of the bubble in effect converts the bubble into a torus, a smoke-ring-like vortex. As a result, the bubble expands, yet loses buoyancy, as it rises.

In the 1950s, Ludlam and American meteorologist Joanne Simpson filmed clouds developing in the tropics. They also wrote formulas for calculating how fast air rises in a cloud. For their mathematical model, Ludlam and Simpson considered updrafts giant "bubbles" of air that erode through entrainment. Each successive bubble warms and saturates the atmosphere a little higher than the previous bubble, making it easier for the next bubble to climb farther without dilution. The cloud thus builds

A broad cumulonimbus tower and its anvil (left) show mature thunderstorm form, while getting new competition from a fast-growing storm tower that's nearly ready to spread its own anvil. NOAA photo.

upward bubble by bubble. The model marked a new maturity of cloud study: it yielded reasonable updraft rates of 7 or 8 m.p.h. Without entrainment, classical parcel calculations with the same tropical data had predicted unlikely 40-m.p.h. updrafts in relatively mild cumulus.

Simpson and her colleagues found other significant facts of cumulus life during their flights in the tropics. They discovered shafts of open air below them inside the growing cumulus. The gaps meant that the clouds were numerous eddies of moist air working as a team—not large, monolithic updrafts. Though the updrafts tossed their small planes up and down with nauseating jolts, the scientists had the presence of mind to photograph isolated cumulus merging into bigger and bigger clouds, eventually forming thunderstorms. Scientists at the time were just beginning to confirm that such teamwork of updrafts happens on an even grander scale in thunderstorms. This work took place in a landmark project in the United States.

For two summers—1946 in Florida and 1947 in Ohio—the Weather Bureau, the armed forces, the aviation industry, and the University of Chicago gang tackled the subject of most appalling ignorance in meteorology. They tracked thunderstorms with every new piece of war surplus equipment they could get their hands on. In all, 200 people took part in the Thunderstorm Project, and it took 22 railroad freight cars to haul the gear from Florida to Ohio between summers. This team spirit (of meteorologists, not updrafts) continues to this day. The Thunderstorm Project of 1946 and 1947 is the legendary progenitor of countless national and international efforts in meteorology that have filled the scientific journals of the last half century with some inscrutable acronyms—even acronyms embedded in acronyms—GATE, GARP, BOMEX, TOGA-COARE, and more. A cooperative spirit in meteorology prevails because scientists have found that by sharing they can obtain satellite platforms, airplanes, ships, radar, supercomputers, and other equipment needed for various specialized experiments. The logistics can get tricky, but the rewards multiply.

Scientists also planned the Thunderstorm Project in part to employ the suddenly idle veteran pilots eager for postwar adventure. The pilots flew P-61 "Black Widow" night fighters back and forth into thunderstorms to take samplings of temperature, humidity, pressure, electrical fields, and winds. Five planes penetrated the clouds simultaneously at different altitudes. In all, the Black Widows flew 1,363 sorties through conditions far more dangerous than many assignments over the European Theater in World War II. The updrafts rocked the small planes unmercifully. The shearing downdrafts, pilots would later learn, can shoot down a jetliner.

Meanwhile, on the ground, scientists organized dense networks of 55 meteorological stations spaced a mile apart. The instruments recorded meteorological factors at one-minute intervals. Six balloon launchers sent other instruments simultaneously into passing storms to measure conditions aloft. The balloons were the new "rawinsondes" that give scientists detailed upper wind information. The system relies on radar, then coming into its own in meteorology. Scientists first used radio to "detect and range" in the atmosphere in 1925 when they aimed the waves heavenward, bounced them off trails of ions behind meteors, and confirmed the existence of the ionosphere at 18 miles up. During the war, air defense officers nervously watched rain showers fill their radar scopes and prayed no enemy planes lurked behind the weather. Now, one person's clutter was another's treasure: meteorologists could finally peer into the clouds. During a storm, the Thunderstorm Project scientists aimed the radio waves vertically to get cross sections of cloud activity and horizontally to get a plan view. The cross sections quickly confirmed that rain can fall without first forming as ice, thus settling a hot debate in cloud physics.

The Black Widow pilots discovered that updrafts are not uniform inside thunderstorms. Belts of calm, cloudy air surround the intense turbulence. Similarly, volunteers from the Soaring Society of America found their gliders spun out of control between updrafts and downdrafts in the storms. The job of investigating these reports fell to Roscoe Braham Jr., one of the project's principals from the University of Chicago (Horace Byers was the other). Braham saw that the turbulent patches were separate convection cells in the storm. Eventually these patches merged into larger cells, a process the scientists caught on radar. Byers later compared this expansion of cells to the multiplication of bacteria. A Florida storm might typically cover 100 to 150 square miles and inside have six separate convection cells, each three to seven miles across. The smooth belts between the cylindrical cells might be less than a mile wide, separating distinct shafts of heavy rainfall. A cell might contain only an updraft, only a downdraft, or both. Byers and Braham soon realized that this finding was essential: not

A thunderstorm over Scottsdale, Arizona, shows lightning flashes to the ground, to the clear sky, and within the clouds. Only a sixth of lightning flashes ever reach the ground to recharge our planet's surface. Christopher Reith/*Weatherwise* photo contest.

all cells in the storm are the same age. A single cell might last only 15 or 20 minutes, but a storm can last an hour as a cooperating collection of young, mature, and decaying cells.

Just like the eddies and updrafts that combine to make a cumulus cloud, the cumulus themselves merge to make a rapidly growing cloud, called *cumulus congestus*. This cloud is several miles wide, though not usually taller than 15,000 feet to start. The cumulus congestus is at first almost entirely updraft—with air ascending at anywhere from a few miles per hour to maybe 60 or 70 m.p.h. The air around it settles gently to compensate. Ice typically forms at the top of the cumulus congestus, and raindrops develop about 15 minutes after the cloud forms. These fall through the weakest and oldest updrafts. The rain makes the air slightly denser than before, dense enough to drag the formerly buoyant updraft downward. This is what Byers and Braham called thunderstorm maturity: coexisting updrafts and downdrafts. The rainy downdrafts entrain drier outside air which evaporates some drops before they hit the ground. If the storm forms in dry air, like many High Plains thunderstorms, the rain and snow might fall out of a base at 10,000 feet and never reach the ground, instead forming a shaft of evaporating moisture called virga. Either way, precipitation or no, the evaporation cools the downdraft, accelerating its fall. The cloud is now a full-blown cumulonimbus.

To a casual observer of storms, the updrafts, not the downdrafts, might seem to herald the mature stage. This is because the most successful updrafts simply crown the storm with an icy cirrus anvil. This thunderstorm trademark reveals an important boundary in the atmosphere. Here, likely anywhere from 30,000 to 60,000 feet, the relatively warm tropopause and stratosphere (where air gets warmer with height) stop the storm's upward growth. A few overeager updrafts may punch their way several thousand feet into the stratosphere, but basically the anvil is the top of the storm. The bottom of the anvil is too light for the troposphere, but the top of the anvil is too heavy for the stratosphere, so the anvil crystals have nowhere to go. Over 10 or 20 minutes, the anvil gets squeezed into a shallower and wider space and spreads out. With fast upper-level winds, the anvil often races many miles ahead of the storm. Sometimes the shade of the anvil cools the ground ahead so much that it kills the storm's chance for survival.

When a storm begins forming ice crystals, chances are it is electrifying as well. Within 20 minutes, lightning and thunder might begin, and the lightning might last several minutes after rain ends. The energy used by the storm to make lightning is almost trivial compared to the huge expenditures of heat used to make updrafts. Even so, the cloud must make about 100 tons of rain for every bolt of lightning.

The cloud generates these sparks the same way you generate them by pulling off a wool hat on a dry day—by separating enough charge to tempt electricity to leap through thin air. Leaping through thin air is not a trivial feat. For an invisible substance, air is actually a pretty effective insulator. We know this because lightning won't strike on a clear, stormless day, even though the atmosphere constantly has a significant electrical potential gradient. The Earth's surface is routinely negative while the atmosphere is positive. For years, scientists puzzled over the negative charge of the Earth's surface. Some thought the charge might be left over from the formation of the planet. Then, in 1899, C.T.R. Wilson and two German scientists showed that the solution to the puzzle was right above their heads. They discovered omnipresent ions in the atmosphere. These charged particles constantly conduct charge away from the Earth's surface. With ions in the air, the Earth's negative charge should drain away in less than an hour. Wilson was convinced that thunderstorms act like a giant battery to constantly restore the charge. He was right: at any given moment, more than 1,000 thunderstorms are raging around the globe; a NASA satellite launched in 1995 to detect lightning shows perhaps 40 to 50 flashes a second. At peak, each lightning flash has a current near 30,000 amps, though some bolts have been measured at 10 times that.

On a fair day, the routine charge separation yields a potential gradient of about 100 volts per meter. Air won't break down for lightning until the gradient is closer to 100,000 volts per meter. When Wilson began to look into thunderstorm charges in the early 1900s, he found positive charges accumulating at the top of the cloud and negative charges accumulating at the bottom. The indirect measurements of storm charges from a distance were so crude, however, that another leading lightning expert, the Englishman George Simpson, found the opposite: positive charges on the bottom and negative charges on top. As a result, one of the few things scientists could agree on before the 1930s was that, somehow, growing raindrops help charge the clouds. Finally, Simpson asked scientists at Kew Observatory to make a small electric charge meter usable with a balloon. More than 100 were launched into storms before World War II, and they seemed to vindicate Wilson's positive-top measurements. But the balloon measurements showed something else significant: the charged regions of the clouds generally hovered far above the freezing level. Whatever caused electrification probably involved ice. Simpson was surprised: "We took it for granted that rain was the chief agent in the generation of the electricity of thunderstorms."

One proposed charging mechanism involved the fair-weather electrical field. This field induces charges on crystals and droplets. Since the ground is negative and the top of the atmosphere is positive, the droplets in between respond with positive bottoms and negative tops. As heavier drops collide with smaller ones, tops rub bottoms and they exchange charges. The big particles end up with more negative charge; the smaller ones rise to the top of the cloud with positive charge. This induction mechanism worked well with the charge separations Wilson observed, and scientists began using it in computer models of clouds in the early 1970s, in part because it was simple to program into the machines then available.

At the same time, debate heated up over an alternative charging mechanism formulated by Bernard Vonnegut, the famous General Electric scientist, and Charles Moore, of New Mexico Tech. They said that charging occurs because of convection, rather than precipitation, thus zeroing in on the defining characteristic of thunderstorms. Instead of saying the rain causes the lightning, they insisted things are roughly the other way around. Positive charges congregate in low-level air, attracted to the Earth's negative surface. Updrafts then lift the positive charges into growing cumulus clouds. At the top of the clouds, these positive charges attract negative ions created by cosmic rays. The negative charges attach to crystals and get cycled by downdrafts toward lower reaches of the cloud.

Vonnegut and Moore did much of their testing in New Mexico. The state has been one of the centers of lightning research since the 1930s, when Everly John Workman started unusual field studies of thunderstorms. The state's desert, with its open spaces and its clear air, offered Workman a unique opportunity to pursue building clouds. His observations with a stopwatch and theodolite (to measure cloud-top altitudes) in the 1930s helped establish that lightning was related to clouds tall enough to host ice crystals. Workman also rigged a Packard with a lightning-proof steel top to mount electric field meters and drove after the storms across the desert floor. It was a good technique, except that Workman became notorious for reading instruments, not watching the road, while he was at the wheel.

Workman established New Mexico Tech's mountaintop weather program near Socorro, in 1956. Here, at 10,900 feet elevation—from an observatory named in honor of Irving Langmuir—Moore and Vonnegut had an ideal setup to establish their unorthodox ideas. They showed, for instance, that the traditional model of thunderstorms needed some revising. Critics of their charging theory had pointed out that young storm clouds don't have the downdrafts needed to circulate ions. So Vonnegut and Moore found downdrafts in young clouds, suggesting a need to modify the thunderstorm growth model. The scientists also began generating positive charges to release in updrafts. In 1984 they generated positive charges with a mile-long, high-voltage wire strung across a canyon beside the observatory. The storms overhead reversed polarity with polarity of the wire, suggesting that convection could possibly charge the clouds.

At Langmuir Lab scientists also had a knack for drawing the lightning to them, rather than chasing it across the desert. For a while they launched charged balloons trailing wires into the storms to trigger lightning. Unfortunately, the balloons burst when hit by lightning, and the wires remained untouched. Eventually the meteorologists adopted a French technique: firing missiles into the clouds to trigger the lightning. Triggered lightning gives scientists a chance to measure currents directly. In the 1930s and '40s General Electric scientists pioneered this approach by using the Empire State Building. The skyscraper gets struck by lightning about two dozen times a year—in one storm during the study it got hit eight times in 24 minutes. The use of rockets to trigger lightning continues to this day. Some scientists are hoping to develop lasers for the same purpose. Because lasers create ionized channels in air, they might divert lightning away from vulnerable power lines. A 24-hour blackout in New York City caused by lightning in July 1977 cost close to $1 billion in looting, so such research has immense practical value.

While Vonnegut and Moore were able to stir up a controversy in a field still shrouded in uncertainty, new observations in the past three decades have put extra demands on lightning theorists. Some of these observations make both the convection and induction mechanisms seem insufficient or unsatisfactory. One finding shows that the negative charges in the cloud concentrate in pancakelike fashion at altitudes where the temperature is close to 5 degrees Fahrenheit. Below it is another positive region, near the cloud base. The thunderstorm is a sandwich of charges, rather than a simple dipole battery. The newer picture of storms suggests that charging may largely be due to collisions of small, soft ice pellets (called graupel) and ice crystals.

Aware that frozen particles might play a part in lightning, scientists in the 1950s and '60s strove to understand the charging tendencies of small

Lightning leaves an indelible mark where it blazed through the bark of a tree. A few inches deeper and the tree would probably have splintered into thousands of pieces from the intense heat. Walter W. Piroth/*Weatherwise* photo contest.

ice crystals as well as large hail. Some scientists sought the extreme cold of Yellowstone National Park, for example, to test collisions among fine ice crystals in the air. It wasn't until laboratory experiments with artificial hail in the late 1970s, however, that cloud physicists stumbled onto the surprising significance of 5 degrees. The scientists grew the "hailstones" by bombarding little ice-coated metal spheres with supercooled mist. Then they measured the charge of their handiwork. Unexpectedly, at temperatures not far below freezing, the hailstones collected a positive charge. Eventually the scientists isolated a peculiar threshold for collisions between small crystals and graupel. When the temperature is below about 5 degrees, graupel emerges with a negative charge and the smaller crystals take the balance of the positive charge. At temperatures above 5 degrees, however, the graupel becomes positive. This mechanism seems to explain the negative charge pancake. The pancake is where the negative, relatively cold graupel collects: crystals above it are positive, and so is the graupel falling below it. No one knows precisely why this charge reversal temperature threshold exists. Recent radar studies in Florida—the nation's lightning epicenter—show that the negative charge collects as expected near 5 degrees, but that the positive charge keeps spreading upward with the updrafts to altitudes as cold as −40 degrees or even −70. The positive spread is probably due to the rising small crystals.

Since charges separate within the storm cloud, most atmospheric sparks fly between the pools of charge within the cloud. The average passenger jet will get caught in the crossfire once or twice a year, drawing the charge to itself. This once was a serious problem: in 1963 a Boeing 707 over Elkton, Maryland, was struck by lightning, igniting a wing fuel tank. All 82 people aboard died in the crash. Since then, commercial jets have improved means of conducting the sudden current on their metal skin.

Only about one out of six lightning bolts is the familiar cloud-to-ground variety that kills more than 100 Americans a year. To call the bolts cloud-to-ground lightning, however, is an oversimplification. Like the air in the thunderstorm, lightning is an up-and-down process. We know this largely because Sir Charles Boys developed a camera in 1902 for taking high-speed photos of lightning. In Boys' camera, two lenses rotate in opposite directions about 25 times a second over a stationary film canister. The lenses distort in opposite directions, thus isolating the up-and-down motion and timing of the lightning strike in detail. Ironically, Boys himself wasn't able to get a good image out of the camera, but in the 1930s Sir Basil Schonland, a South African educated in England, took the Boys camera to the dark nights of the Transvaal province. After several frustrating years with rain- and hail-smudged images of bright streaks, Schonland made more than 150 photos of lightning that resolved it into its components.

Schonland showed that an extremely faint stepped leader of charge descends from the cloud. About 95 percent of the time, this leader comes from the negative charge pool. A fringe of positive charge at the bottom of the cloud may be the initial target, but the negative channel overshoots the cloud base and heads toward the ground. The channel of negative ions lurches along at about 200,000 m.p.h. in roughly 150-foot leaps, rests, and then leaps again maybe 50 millionths of a second later. A seemingly random staircase, the leader—a nearly invisible channel—zigs and zags its way downward, ionizing air molecules as it goes. Within perhaps 100 feet of the ground, it draws a positive response from the Earth. Positive? Yes, the surface of the Earth may be negative overall, but it has a relatively unlimited supply of both negative and positive ions. The cloud typically casts a "shadow" of positive charge on the ground. Often, near an electrical storm, pointed objects like ships' masts, trees, and flag poles actually discharge these ions in a luminous phenomenon called St. Elmo's Fire. G.E. scientists found that a really tall pointed object, like the Empire State Building, actually initiates its own upward stepped leader, making an upside-down lightning of sorts. You can tell with a naked eye if lightning is initiated from the ground this way: if the path branches, it branches upward like a tree.

The upward connection is called a return stroke. But it would be a mistake to think ions are racing up the leader channel. The current peak, not the electrons, zips upward. The negative ions (electrons, really) at the bottom of the leader descend the newly formed channel first, then ions farther and farther up the channel move downward. The peak of current progresses up the stepped leader at 60,000 m.p.h., and the channel glows with the brightness of a 100 million light bulbs. The light lasts only as long as the current. For a return stroke that can be anywhere from a few thousandths of a second to an occasional, damaging sustained stroke of more than 1/10th of a second. After a few hundredths of a second respite, many lightning flashes repeat the exchange: first a downward leader, then a return stroke, followed by a gap barely long enough for the human eye to detect. As a result, the lightning appears to flicker. Scientists once recorded 47 return strokes in one lightning flash. Each return stroke heats the air in the ionized channel suddenly, and the expansion sends out a shock wave we hear as thunder.

About one in 20 bolts actually lowers positive charge to the ground. These bolts are among the most powerful, and foresters especially watch for them on National Lightning Detection Network displays, since positive bolts may be responsible for most lightning-triggered forest fires. That's about 1,500 fires each year in the United States. The positive lightning often seems to emanate from high in the cloud, which is not surprising since positive charges congregate at these altitudes.

None of this explains the peculiarity of lightning when it strikes. The lightning channel is only about as thick as the shaft of one of those golf clubs it seems to love to strike. But in that tiny zone of air the temperature nearly instantly jumps to about 54,000 degrees Fahrenheit—six times the temperature of the sun. By comparison, iron melts at only 2,700 degrees. Should the lightning travel through a tree trunk, this heat vaporizes the wood in a pencil-thin channel, generating such intense steam pressure that the whole trunk is likely to explode. If the lightning passes along the bark, on the other hand, the tree might split open yet survive.

People, too, can survive lightning. Maybe less than one in 10 of the people struck by lightning dies from it. They don't all even suffer burns, despite the temperatures. Frequently, wet clothing conducts the electricity to the ground, not skin. But there is little guarantee of safety around lightning. Golfers in Florida are often struck by lightning after the sky clears up, simply because the thunderstorm passing over the horizon can still strike from five or 10 miles away. There is no good way to predict where the lightning will strike, or whether it will jump near the ground. In Chicago Heights, Illinois, in July 1987, spectators at a basketball tournament in Martin Luther King Park headed for the gate when the games were rained out. Lightning struck a tree "like a big ball of fire" near the exit, and instead of streaming harmlessly into the ground, the current jumped onto the chain-link fence around the courts. Sparks flew, injuring 30 people nearly simultaneously as the bolt ripped through the crowd. One survivor remembered seeing the light charge at her face, then woke up five feet away from where she had stood.

Chicagoans got another taste of the bizarre a few years later. In the suburb of Crystal Lake during a snowstorm, workers changing shifts at a machine shop trudged out to shovel their cars free. Then lightning hit nearby with a deafening blast of thunder, and the current surged through the snow in the parking lot. Eleven men were hospitalized with temporary paralysis and other pains from shock. Oddly, a man riding a snow blower in the parking lot was uninjured. And just to prove that the Windy City is indeed a center of bizarre electrical phenomena, a flash of lightning in Chicago entered a house through a cable TV line one night in 1991. It ignited a bed where someone was sleeping. The victim bolted out of bed quickly, suffering only shock.

Lightning is the defining pyrotechnic display of thunderstorms, but the moment charging begins with the appearance of ice is also often the beginning of hail. Hail starts when tiny ice crystals gather supercooled water and grow into graupel. The milky-white, crunchy graupel pellets are aerodynamic spheres and cones an eighth or even close to a quarter of an inch in diameter. Raindrops, by contrast, can't get this big because they break up as they spread into unstable hamburger-shaped blobs (nothing like the teardrops you sometimes see in drawings). As big as it is, graupel falls too slowly, at less than 10 m.p.h.; in strong thunderstorms it rockets skyward. But tiny water droplets rise even faster, so the graupel becomes an embryonic hailstone by dominating the cloud particles around it in the updraft. It grows by colliding with the light droplets and begins to fulfill its destiny as a large hailstone. (Sometimes frozen raindrops also serve as hail embryos in wetter parts of the storm.) When the ice pellets reach about a quarter of an inch in diameter they are officially hail—though at this point only pea-size hail.

Most hail larger than a marble acquires ice in several distinct layers (the record is 25). To examine the layers, meteorologists began autopsying hailstones in the 1950s. They cut the stones into sections not much thicker than a human hair and examined them under microscopes backlit by polarized light. The special light revealed the differences in the way the ice accumulated on the embryo. Generally, each layer might form in one of three ways: First, the supercooled droplets might hit the stone and spread thin before freezing. This type of layer has extremely clear and dense ice with relatively few crystals. Second, the droplets might hit the stone surface but not spread out. In this case the stone fails to relinquish the latent heat released during freezing; some water remains liquid, trapping air bubbles in an opaque, spongy layer. Third, the supercooled droplets might freeze nearly instantly. This type of layer traps air easily and is not very dense. Scientists are interested in the layers in part because they signify the temperatures high in the storm: the more bubbles in the ice, the colder the conditions—in part because colder water dissolves air more readily.

The layers of ice give the impression that the hailstone rides a roller coaster of air currents just like Lieutenant Colonel Rankin did after bailing out of his fighter. Each ascent could produce a new layer. It is easy to see how a mature, multicell thunderstorm can create this roller coaster for hailstone growth. The hail is very sensitive to updraft speeds as it grows, because the larger it is, the faster it falls. (It's a good thing Galileo didn't try to demonstrate gravity by dropping hailstones from the Leaning Tower of Pisa—with varied densities and aerodynamic properties, they'd have ruined the experiment.) If the ice rides the wrong updraft at the wrong time, no hail forms. A crystal too small for its updraft will rise so fast that it won't get the benefit of the 10 or 15 minutes of slow accretion it needs to reach graupel size. And graupel in a weak updraft will merely fall to the ground prematurely. But a multicell storm often proves a good hail parent, since at any given moment, somewhere in the storm cloud, there is a cell with the proper updraft strength. At first, the tiny ice crystals might grow in a rather weak updraft. This gives them time to rime and to clump together into graupel, which contains about 1 million cloud droplets. The graupel will then tumble out of a weak cell and get caught in a younger cell nearby and remain aloft as it grows heavier and becomes a hail embryo. In a still stronger updraft, the embryos can grow even larger. By the time the hail is golf ball size, it has collected about 10 billion supercooled droplets.

Given the extreme moisture and updrafts in some thunderstorms, practically any solid body would seem able to grow into a hailstone. Five glider pilots bailed out in a thunderstorm over the Rhön Mountains of Germany in 1930. The supercooled water encased them in ice. Miraculously, one survived. At other times, icy clumps fall from the sky entombing small fish, insects, and even gopher turtles that somehow get into the storm. The largest authenticated hailstone of purely atmospheric provenance fell in Coffeyville, Kansas, on September 3, 1970. It had a diameter of 5.5 inches and weighed one and two-thirds pounds. At this size, hail is quite deadly. A 1986 storm with grapefruit-size hail killed 92 people in Bangladesh. The worst hailstorm on record, in India, on April 30, 1888, killed 246 people, most of them were knocked unconscious first and then suffocated and frozen under icy drifts. Within four days in July 1953, two hailstorms in Alberta, Canada, killed more than 64,000 wild birds.

The deadly double jeopardy of a thunderstorm is that bigger hail falls faster than small hail. A marble-size hailstone will plunk onto your head at just under 40 m.p.h. Baseball-size hail zips along like a Nolan Ryan fastball—close to 100 m.p.h. That's fatal if you don't wear a helmet. These generic speeds were established by scientists as early as the 1930s, when they tested the falling rate of models of balls of ice. Since then, meteorologists have come to appreciate the importance of different shapes of hailstones. Small hailstones tend to be spherical, but the ultimate stone shape is sensitive in part to the embryo shape. It is also sensitive to the concentration of water in its environment. As it grows, a hailstone that doesn't rock and spin too much in the wind can start to look like a space capsule. These cone-shaped hailstones rarely exceed an inch in diameter.

Larger stones start to exaggerate a lumpiness, or lobe structure, that originates early in the stone's growth. Scientists have actually grown such lobed hailstones in wind tunnel tests using highly concentrated mists of very small droplets. The lobes help the stone grow large, because the increase in surface area helps the ice get rid of latent heat released by freezing droplets. Melting from latent heat release might explain why some hailstones are spiked like a punk-rocker's hair. The spikes might form the way ordinary icicles do on your eaves—by water flowing around the ice.

In one wind tunnel test, scientists at SUNY Albany grew an inch-long spike on a hailstone. They blasted the artificial stone with supercooled droplets and found that the latent heat release at freezing was just enough to melt some of the outer ice, making a film of water that flowed to the back of the stone, out of the wind. There it refroze on an old bump of ice, gradually building the icicle. To test the fall speeds of different shapes of hail, scientists made plastic replicas and dropped them down a mine shaft more than a mile deep.

Such giant, strange hail does not necessarily have to travel up and down like a Yo-Yo in the storm to grow sufficiently. Some radar studies of thunderclouds show that water collects in a big supercooled bath perhaps 15,000 to 30,000 feet high. Such observations prompted Soviet scientists to develop a new model of hail growth. They said the stones could grow very large with just one ascent in a storm. They noted that as soon as super-cooled water drops approach a quarter of an inch in diameter, their terminal velocity stabilizes around 20 m.p.h. Further growth doesn't increase that speed much, given the terrible aerodynamics of the hamburger-shaped drops. Meanwhile, thunderstorm updrafts tend to accelerate to a peak somewhere high in the storm, then decelerate quickly at the top. So drops grow as they rise in the storm but never fall out because their terminal velocity never exceeds the updraft speed. They simply stop rising when they reach the top of the updraft where air decelerates. Where the updraft slows to 20 m.p.h., many large, probably supercooled drops accumulate but remain stranded. The Soviet scientists believed that hail grows quickly in this cloudy reservoir as soon as an ice crystal forms.

The Soviets used this model with some success, beginning in the 1960s. With radar, they aimed 70-mm howitzers at the watery region of the clouds and blasted away. This was an ancient hail mitigation strategy based on the principle that sudden noises might shatter the hailstones. The technique was used with moderate success in 1966 over the Kericho highlands of Kenya, near Lake Victoria, where tea plantations suffer more hail than any other place in the world—on average more than 100 days a year. (By contrast, Cheyenne, Wyoming, the most hail-prone U.S. city, gets about 10 days of hail a year.) The Soviets used their guns scientifically, however. The shells were timed to burst at about the 20-degree Fahrenheit level, spreading a mixture of gunpowder and silver iodide. The silver iodide overseeded the cloud, and large numbers of crystals formed. Given the same amount of water in the cloud, more crystals means more hail-stones, and each hailstone is smaller, less damaging.

In the 1970s American and Canadian scientists tried to verify the Soviet hail model in Plains thunderstorms. Scientists at the South Dakota School of Mines and Technology had recently developed the perfect tool to investigate the hail-making regions of the clouds. They outfitted a Navy surplus T-28 trainer for the job. The T-28 is no ordinary airplane. The 1940s vintage craft is airworthy enough to take the worst a beginning pilot can throw at it. But in order to survive the hail environment, the scientists added 700 pounds of armor to the tiny single-engine plane, which is also loaded with redundant meteorological and navigational instruments. The engine is powerful enough to thrust the plane out of the storm with as much as an inch-thick coating of ice on the wings. The scientists deemed this more practical than outfitting the plane to deice itself: it would be nearly fruitless in the thick environment of supercooled water. Despite all precautions, pilots have found the T-28 engine occasionally clogs with ice and shuts down in mid-air, but so far, after 900 successful sorties into storms, it always restarts when necessary. Lightning also frequently pits the armor surface. And even with the armor, meteorologists use ground-based radar to guide the plane away from hail larger than an inch in diameter.

Scientists sent the T-28 high into the clouds to count water and crystal concentrations and measure updrafts. The storms seemed to generate high concentrations of crystals on their own, which meant that the Soviet-style hail seeding would seemingly be unnecessary in many American storms. This also meant that the high-altitude pool of supercooled water wasn't materializing the way the Soviets described. Apparently another process makes most large hail in North America. At this point, scientists knew they had to turn to a different kind of thunderstorm to explain how updrafts made large hail. This had to be a storm quite different from the traditional multicell storm described in the late 1940s, a storm that could allow downdrafts and updrafts to coexist for a long time.

In the Thunderstorm Project, meteorologists occasionally detected downdrafts when their anemometers spun to opposite directions as cold, sinking air spread out on the ground. In a matter of minutes, the downdraft air spread out several miles ahead of the storm's rainshaft, creating a miniature front several thousand feet thick. The leading edge of the downdraft became known as a "pressure jump line," because the cold down draft pressed down on the surface, creating a small high-pressure area. Though meteorologists had long recorded tiny pressure increases under thunderstorms, the Thunderstorm Project confirmed the structure and significance of this small-scale, frontlike feature. This gust front feeds the storm by kicking up fresh ascending air that will form new updrafts if conditions are appropriately unstable.

Mammatus clouds sprout from the underside of a severe thunderstorm's anvil. Once thought to be a precursor of tornadoes, these strange shapes form from gentle unrest between cloudy and clear air. Cherry Russell/*Weatherwise* photo contest.

The downdraft thus can help to maintain the storm. Byers and Braham chided a century's worth of meteorologists for obsessing over the updraft and neglecting the downdraft. "One may say that heretofore meteorologists ... have been barking up the wrong tree," Byers wrote. "The downdraft is by far the most striking feature, at least ... at or near the ground." It was a bit galling to Byers, however, when in 1951 he received an astounding letter from a young Japanese college physics instructor. Until then Tetsuya Fujita had lived most of his life in Nakasone, a tiny rice-farming village on the volcanic island of Kyushu. He had never heard of the Thunderstorm Project, yet working alone, he discovered the importance of downdrafts, diverging surface winds, and pressure jump lines under thunderstorms. "We spent millions to discover downdrafts," Byers admitted, and here one man spent a few yen and essentially equaled the feat. With a little help from Byers, Fujita fulfilled this promising start, becoming one of the key figures in understanding severe weather.

Born in 1920, Fujita graduated from Meiji Technical College in Kyushu during World War II and immediately joined the staff of the college, rising quickly to junior faculty rank. Fascinated by geology, Fujita developed his legendary artistic skills by drafting topographical maps. Mostly, however, he studied mechanical engineering. In 1945, with the war reaching the desperation point, the Japanese government sent Fujita's students to work in coal mines. Undeterred, the young instructor brought a barometer and thermometer with him to check the mine shaft depths against topographical charts. Fujita's interest in meteorology grew as he began to find pressure variations underground that were unaffected by atmospheric depth.

One night in 1947, a terrific thunderstorm broke out over northern Kyushu. Fujita decided to map the storm as it passed. The enterprising physics teacher grabbed a watch, a pencil, and a pad and recorded the direction and time of each lightning stroke. After examining local weather station data, he concluded that the lightning was clustered around several tiny, moving, high-pressure areas. Then he applied what he learned in the mines about surface pressures that aren't in pure balance with atmospheric depth and gravity. He speculated that each high was probably a downdraft pressing on a small spot of the Earth's surface. He calculated the unobserved wind divergences likely caused by the downdrafts.

Fujita didn't find anything like these discoveries in the meteorological literature available to him. He read a paper on the small pressure surges, which he called "thundernoses," to Japanese colleagues. They politely refused to comment. An intensive thunderstorm research program in Japan in 1940 had revealed surface divergence, but no one had thought to blame downdrafts. Fujita continued working alone, pioneering the analysis of tiny pressure variations with topographical precision. Instead of rocks and cliffs, he surveyed frontal pressure patterns. One squall line produced a rare tornado in Kyushu in 1948, and Fujita examined damage patterns in the rice fields and deduced the motion of thundernoses associated with the damage. His work anticipated the American research on severe storms that developed after the Thunderstorm Project.

Fujita's colleagues thanked him for one lecture by giving the budding meteorologist a copy of a monograph fished out of the trash bin of the U.S. radar station on Kyushu. It was "The Nonfrontal Thunderstorm," by Horace Byers. Impressed by the Chicago professor's work, Fujita shopped around for an English-language typewriter and found a used one for about $50 that he later admitted was the most expensive thing he had ever bought up to that time. After months of laborious typing, Fujita began sending his papers to Byers for comment. Byers was deeply impressed, and after Fujita obtained a doctorate in meteorology in 1953, his mentors in Tokyo sent Fujita to Chicago for further work. Byers arranged for Fujita to join the faculty there, a move that changed the face of thunderstorm research. Fujita (who became known by the name Ted in his adopted country) and Morris Tepper of the Weather Bureau looked closely into the downdraft-produced boundary ahead of the thunderstorms. Together the scientists pioneered a whole new field, called mesometeorology—the analysis of small systems of weather like thunderstorms that often slip unnoticed between standard weather station networks. (Today "mesoscale" refers to almost everything between giant "synoptic scale" extratropical storms and the compact whirls of tornadoes.)

A quarter century later, Byers' emphasis on downdrafts would gain new meaning in one of the biggest breakthrough storms of Fujita's remarkable career. At first lightning caught the eye of witnesses near the awful scene just northeast of New York's Kennedy International Airport on the afternoon of June 24, 1975. Pedestrians hurried to shelter as thunderstorm rain cascaded onto the streets. The lightning drew the manager of a nearby tennis club to the window to watch the planes approaching low under the 3,000-foot-high cloud base. At that moment she saw a blue-and-white Boeing 727—it was way too low, maybe a few dozen feet above the streets. A policeman nearby saw it, too, and swore a giant ball of lightning coursed through the Eastern Airlines flight 66 at that last moment as it passed over Rockaway Boulevard in Queens.

The pilot, Captain John Kleven, turned on the windshield wipers as the plane entered the thunderstorm's rainshaft at an altitude of 700 feet two miles from the runway. A steady head wind favored the plane, and Kleven cut power back. Suddenly, with a mile to go, the rain seemed to weigh the plane down. The plane rocked and lurched in the wind, which stomped on it from above. The right wing dipped low and the jet veered off its landing path. It sank quickly to below 50 feet, swiping a tower of tracking lights leading it to the runway. Kleven desperately swung the plane to the left but it was too late. An explosion ripped through the aircraft as it plowed through more airport lights, flopped onto Rockaway Boulevard, and skidded into a field. No one who was watching could miss it; the plane was consumed by a ball of flame that leaped several hundred feet into the air. Only 11 of the 124 people aboard survived, most of them in the tail section of the plane.

Accident investigators knew lightning was an unlikely culprit; they suspected wind shear. The pilot of a KLM 747 that landed 11 minutes before Eastern 66 said that his plane experienced sharp wind shear at an altitude of 300 feet and that he had applied power suddenly to compensate for a strong tail wind. Three minutes later another Eastern jet approached. Just over a mile from the runway, at an altitude of only 300 feet, a stiff head wind suddenly increased air speed even as it pushed down on the plane. Seconds later, the downdraft suddenly turned into a tail wind with less than half a mile to go. In danger of losing air speed too quickly at less than 100 feet in altitude, the pilot applied full thrust and managed to escape. Not wishing to tempt fate, he decided to land at Newark, instead. But the wind shear evidence was confusing: several other pilots landing on the same runway at JFK didn't experience any severe winds in the rain.

Eastern Airlines officials sent flight data to Fujita in hopes that he might explain the quirky circumstances of the crash. Fujita's expertise in ferreting out the smallest details in thunderstorms proved critical. He found evidence of fast-spreading winds at the ground. These star bursts of damage were tightly organized—between a quarter of a mile and three miles wide. Nothing of this small size had been noted from thunderstorms before, but Fujita, who during his career had flown thousands of miles over tornado damage tracks, knew he had seen the pattern before. He studied his aerial photographs from previous storms for star bursts and found some in a massive tornado outbreak that had struck the Midwest and South the year before. He remembered where he had first seen such a telltale sign of a downburst of wind. Back in Japan, in 1945, he had surveyed the damage of the atomic bomb blasts in Nagasaki and Hiroshima. The spreading shock waves had leveled much of the damage in a similar radial pattern. Even without seeing the wind, Fujita knew that the Queens thunderstorm had subjected Eastern 66 to a similar fanning out of winds on a much smaller scale.

Knowing what to look for, Fujita ferreted out the tiny downbursts from the flight recorder data for more than a dozen planes that had approached JFK before and after Eastern 66. He proved that the storm's rainy downdraft produced at least three separate bursts of descending air. Each burst spread out at the surface, creating diverging winds up to 400 feet in altitude. Some planes managed to land in the rain in between the downbursts, but a few had got caught. One had crashed.

Some meteorologists immediately attacked Fujita's findings. Downdrafts from thunderstorms were nothing new, they claimed. But Byers, who for more than a quarter century had advised pilots about dangers in downdrafts, agreed with Fujita's contention that these were newly discovered phenomena on a much smaller scale than anything previously studied. Fujita eventually defined a class of intense downdrafts called microbursts that cause damaging winds over an area smaller than two and a half miles across. A study of plane crashes between 1970 and 1985 showed that

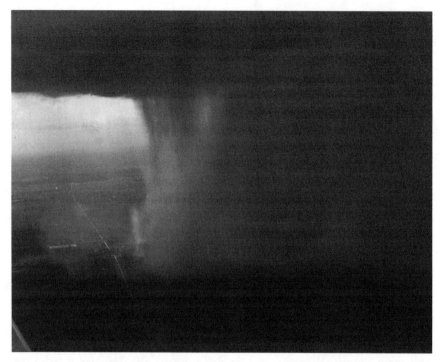

A microburst plunges to the ground and then spreads dust across Illinois fields, spinning up a small whirl along the shearing outflow boundary. Ronald Brewer/*Weatherwise* photo contest.

nearly 600 people had died in microburst accidents. Fujita later got his vindication when a Doppler radar study of microbursts near Denver tracked an average of one a day in a 40-by-40 mile area.

Fujita's microburst studies finally explained to pilots what had been happening to them for many years around thunderstorms. Planes that get caught entering a microburst experience head winds that endanger their approach to a runway. Pilots compensate by pulling back on engine power until a sudden tail wind forces them to scramble to gain air speed. But the downdraft in between often puts the plane at a nose-up position; a last-resort increase in thrust can be effective but can also stall the plane. Transportation authorities used a flight simulator to recreate the Eastern 66 landing conditions and found 80 percent of the pilots crashed in the test.

The entry of dry air at midlevels of the storm is essential to developing a microburst. The dry air evaporates rain falling from above, and the cooling caused by evaporation creates a large bomb of cold air that barrels down toward the Earth along with the remaining precipitation. The cold air downdraft lasts only about 10 minutes or less. When it hits the ground, the descending air spreads out fast, and with more air coming down behind it, the horizontal winds generated may keep accelerating for a few minutes. These winds are not just dangerous for airplanes. Fujita's aerial surveys of suspected tornadoes have shown that people can mistake microburst damage for twister damage. Microbursts uproot trees and flatten cornfields. One microburst struck near Andrews Air Force Base in Maryland just minutes after Air Force One touched down with President Reagan aboard. Surface gusts hit 149 m.p.h. near the runway.

Fujita's study of microbursts near Chicago showed that nearly two-thirds occurred with rain. The peak winds mounted in only a couple of minutes. But in the Colorado study, only one-sixth of the microbursts were wet. In the high-based thunderstorms there, the microbursts originated some 15,000 to 20,000 feet above the Plains. This left plenty of time for the sudden downdrafts to pull in cold dry air around them and evaporate the precipitation. Winds 3,000 feet thick spread out with nary a drop at the ground.

Based on these findings, pilots now have strategies for avoiding microbursts or for pulling out of them if caught inside. But microbursts can pose a wind shear even more complex than the head wind to tail wind shift that downed Eastern flight 66. A Delta wide-body jet approached Dallas–Fort Worth Airport as a thunderstorm intensified near the end of the runway on August 2, 1985. After circling for 20 minutes, the plane began to descend toward the airport and disappeared in the intense rainshaft at about 875 feet. A 25-m.p.h. head wind buffeted the plane in a matter of seconds, but the First Officer, who was flying the plane, knew what to do. He pulled back on the throttle slightly, then coolly increased thrust to maneuver through a 25-m.p.h. downdraft, followed by a 23-m.p.h. tail wind. Several seconds later, however, the plane began to pitch and roll in violent shear. Despite the extra power, the L-1011 careened out of control and plummeted toward the ground.

On Highway 114, at the edge of the airport, traffic slowed to a crawl in the downpour. Anthony Rogers couldn't see more than 30 feet of the road in front of his Camaro when suddenly a giant tire appeared out of the heavens and bounced off his hood. He pulled over. The Toyota in front of him was stopped too. Its roof was sheared off and the driver killed. Aboard the Delta airliner the 163 people aboard felt the sudden landing with a thud and held on as the plane skidded to a stop within 100 yards in the field next to the highway. For a precious last few seconds, they sat in stunned silence. One survivor remembers seeing a flight attendant's face, her eyes wide with terror. The plane exploded and cartwheeled across the field. Most of the 34 survivors sat in the rear, under the jet's big tail engine. From the highway the puzzled Rogers looked over at the airport and saw through the gloom of rain the tragic ball of fire erupt from the plane.

The leading edge circulation of the microburst outflow had downed the airliner. As a microburst spreads, it often curls back on itself as it pushes against the still air around it. The expanding vortex is a bit like a donut of shearing winds, much like the overturning bubble rings cloud physicists once studied. The curl at the edge of the spreading microburst can tower well above 1,500 feet, making small updrafts and downdrafts that can easily confuse pilots. The Delta plane was buffeted by the curl just as it pulled through the microburst tail wind.

Dallas–Fort Worth actually had a crude microburst alarm system standard at the time. It consisted of six anemometers mounted a bit more than a mile apart around the airport. Together they could show surface divergence; they were linked to a computer that alerted controllers to wind shear potential. Unfortunately, the surface-based system couldn't tell anyone what was going on in a thunderstorm a mile away, nor could it give indications of downdrafts aloft until they hit the ground. In this case surface winds initially were only about 10 m.p.h. Twelve minutes after the Delta L1011 crashed, the alarm went off as 80-m.p.h. gusts swept across the tarmac, damaging several planes as baggage carts slammed into them. Another parked plane suffered damage from lightning. Now, more than a decade later, new Doppler radar at airport terminals tracks wind direction and speed at higher altitudes, improving microburst detection at airports.

Downdrafts are deadly to planes but vital to the storm because they kick up new convection cells, prolonging the life of the cloud. The only meteorological solace in the matter is that downdrafts also kill the storm. They eventually choke off the supply of warm air from below. On the ground the storm seems unchanged—high winds and heavy rain may continue—but no updrafts survive the onslaught of precipitation, and the storm is quickly consumed by downdrafts. Within an hour of its first cumulus, the storm may die out completely. Describing this evolution was the crowning jewel of the Thunderstorm Project's achievements. The Thunderstorm Project had yielded an accurate, but limited picture of how garden variety storms work. These storms, often called "air mass" thunderstorms, crop up in isolation in stagnant hot, humid air. The 1940s vintage model remains remarkably current, though the difficulty of modeling multicell storms such as these on a computer shows meteorologists that the mechanisms involved are a bit more complex than once believed.

Despite the success of the Thunderstorm Project, meteorologists knew the results were woefully incomplete considering the wide variety of storms they were seeing elsewhere. In two summers of storm tracking, the Thunderstorm Project scientists found only one storm with updrafts exceeding 60 m.p.h. None spawned tornadoes or serious hail damage. Clearly, a thunderstorm can be much worse. The air mass thunderstorms studied in Ohio and Florida were a good beginning, but squall lines often host more severe weather. Air mass thunderstorms usually don't last long enough to cause serious havoc. Squall line thunderstorms, however, get started along frontal boundaries or in air disturbed near and parallel to fronts. These storms can last many hours, and they move quickly, making them even more dangerous. By ganging together, they can forestall the inevitable decay that visits air mass storms. They can keep the downdrafts from destroying the updrafts. Whereas the individual cells in the line move with midlevel winds, the squall line as a whole shifts semi-independently of winds around it. New cells pop up where outflows from the line storms meet incoming warm air, often southeast of the storms. So the chain of storms begins to add links southeastward, sometimes growing to nearly 1,000 miles long.

A squall line is the most violent wing of a large extratropical weather system. These lines typically evolve at or ahead of the cold front. This is often where moist air races northward in the warm sector of the low-pressure system. In the tropics, where cold fronts don't dare venture, squall lines tend to move west with the prevailing winds after forming near the convergence of currents from the northern and southern hemispheres, near the equator. They lack the punch of a midlatitude squall line in which updrafts (at 50–100 m.p.h.) are about four times as strong. Midlatitude storms get their strength from stronger wind shear with height, strong heating over big continental areas, and even higher cloud bases (due to drier surface air), which give downdrafts more space to accelerate.

Precisely why some thunderstorms band together into a squall line and others remain independent is a mystery of meteorology. But once formed, a squall line can last 5 or 10 hours or, in the case of a powerful low-pressure system, more than a day. The mechanisms for this maintenance are byzantine but, meteorologically speaking, quite beautiful. Much of the line's mechanics depend on the power of its collective downdraft-induced gust front, which races out ahead and shovels up the warm moist air needed for convection. The low-level outflow ahead of the storm some- times twists the storm; viewed on end in cross section, the flow overall rotates around a horizontal axis. In a sense, the sporty new cells tilt their seats back as the line as a whole shifts into high gear. At first, squall line cells are nearly vertical—not a very long-lasting posture once precipitation forms atop them. After an hour or so, they lean back into a comfortable position they can maintain for hours, according to supercomputer models run in the late 1980s. As they lean back, the storms separate downdrafts from updrafts. The rain shaft is directly behind the newly forming up- drafts. An inch of precipitation might fall in a half hour as they pass.

The squall line circulation develops a wide backside. The convective cells may be only a few miles across, but the old updraft air continues rising, more gradually, through a back-spreading anvil of cloud 50 or even 100 miles wide. The storm towers cast ice crystals into this spreading wind. Here in the squall line's broadly rising back 40, the crystals fall into a thick layer of moist stratus cloud beneath the anvil. The stratus reap what the thunderstorms sow: the crystals clump into large snowflakes that begin to fall more steadily toward the ground.

The snow encounters a sharp wind shear near the bottom of the stratus. Here midlevel dry air moves in the opposite direction: it gradually subsides and enters the back of the storm line. This air is pulled into the storm by a midlevel pocket of low pressure (at about 10,000 feet) probably caused by the stretching convective towers, because at the bottom rainy air subsides, yet at the top the anvil rises slowly. The snowflakes from the anvil melt in this rear inflow, yielding a wide area of gentle rainfall. The rear inflow can eventually catch and collide with the gusts ahead of the squall line, which might be moving at 45 m.p.h. This creates a vicious, self- perpetuating cycle: the dry rear inflow—cooled by evaporation—pours onto the ground behind the squalls. In turn, this strengthens the subsi- dence in the rain shaft of the main downdrafts, which can plummet at more than 10 m.p.h. The invigorated downdrafts then stretch the cloudy air more, sucking in stronger rear inflow at midlevels.

In the late 1970s and early '80s, scientists began to recognize from satellite images that thunderstorms don't always flock together in lines. The squall line is merely one type of mesoscale convective system (MCS), as meteorologists call the groupings. These packs of marauding thunder- storms may be triggered in large low-pressure systems, as squall lines usually are. But many of them form in weak cyclonic patterns or all by themselves. Either way, unlike squall lines, they are free to pursue an independent destiny.

Free to choose a shape, they look much different from their squall line cousins. The worst of these groups are the mesoscale convective complexes (MCCs) that often form circles of thunderstorms with a continuous anvil spreading over an area the size of Kansas. A hundred air mass storms could fit inside a well-developed MCC. As large as they are, such storm systems can disappear in the cracks of numerical weather prediction programs. Furthermore, they don't necessarily move with the midlevel winds (at about 10,000 feet) that seem to steer lesser thunderstorms. As the thunderstorms coalesce, they are emboldened to move as they wish.

Afternoon thunderstorms blaze trails of convection through the instability over the prairie. These storms often originate in the clouds that pop up over large mountain ranges like the Rockies and move eastward later in the day. They may be strong enough to produce hail and tornadoes, and they may originally be isolated or organized as lines along atmospheric boundaries. In any case, this daytime convection helps create conditions conducive for a gathering late at night. In the afternoon, the abundant updrafts and condensation warm the air around the storms, not just inside them. This creates a large low-pressure area at middle and upper levels of the troposphere, perhaps 10,000 feet and higher, and the MCC begins forming a giant warm-core rain factory. In the tropics, such areas of warmth and rain gather amid thunderstorms, enhancing the chances that storms will organize into a hurricane. Local storm-making updrafts are no longer buoyant solely because of their own heat and evaporation. The midlevel warmth and moisture pooled in the MCS encourages vigorous, towering convection, which in turn gathers low-level air from farther and farther about, in a sort of chain reaction. In meteorological lingo, MCSs make a convective instability of the "second kind." In the Caribbean this process led to the worst weather disaster of 1998, Hurricane Mitch.

The buoyant midlevel core draws more air into the storm over many miles from all directions. Eventually, if the influx radius is 100 miles or more (a threshold that decreases with greater latitude), the complex runs afoul of the deflecting force of the Earth's rotation. The cloudy mass begins to rotate counterclockwise (in the northern hemisphere), a bit like a hurricane. Above, the outflow from the rising air is similarly widespread, and the resulting upper-level high pressure begins to rotate in the opposite direction. The MCC outflow (reaching temperatures below −112 degrees Fahrenheit) is so tall and strong that at jet stream levels it actually overpowers environmental flow—upper currents must take a detour around this atmospheric road block. Airlines must watch these MCCs carefully. Rerouting a flight on the wrong side of the anvil can mean costly (and bumpy) head winds at the tropopause boundary. Rotating MCCs are notoriously stable, surviving well in marginal convection conditions. The MCCs survive usually 10 or more hours through the night and tend to move only 10–20 m.p.h.—much slower than squall lines.

With this widespread rotation, the thunderstorm towers cautiously circle their wagons around a campfire of convection for the night. Their fate is often to end up as a whirling mass of rainy stratus clouds rather than remain as proud, towering thunderstorms. By nightfall, more than 30,000 square miles of cloud has formed, buffering the thunderstorm towers from the hostile night environment and filling in the gaps with widespread rainfall comparable to that of a hurricane. The cluster takes advantage of a special influx of fuel that peaks in early morning hours. This is the low-level jet, a northward stream of Gulf air that can reach as far north as the Dakotas. The jet accelerates after dark, providing the instability the complexes need. But the rotating maturity of the MCC marks a transition in its character. The thunderstorms that established it now blur into a huge core of widespread, weaker ascent. Eventually, stratiform clouds that spin slower and slower dominate the MCC. In fact, by daybreak some MCCs linger as subtly rotating cloudy sheets at midlevels, where the low-pressure once formed. This stale but warm stratus region then kicks up new thunderstorms for a second day. Eventually these hardy storm formations die out simply by moving on to the Atlantic, well past their unstable spawning grounds.

This transformation ultimately determines the effect of the MCC. Rather than produce many tornadoes or much hail, MCCs are known for soaking communities below with copious rains. A community caught under an MCC can easily have several hours of thunder and lightning, not to mention tornadoes. But the rain is the real problem. An MCC over Johnstown, Pennsylvania, in 1977 showed why. It caused a reprise of America's most famous flood. This one killed 77 people in the town where more than 2,000 died in 1889 after a dam burst. An endless stream of MCSs, many full-blown MCCs, caused the 1993 floods on the Mississippi and Missouri Rivers that cost more than $20 billion in damage.

Dozens of MCCs form each year east of the Rockies during spring and summer. One study showed that these systems make a quarter to half of all the annual rainfall in much of the central Plains. The average MCC drops a half inch of rain over a swath of land equal to the combined areas of

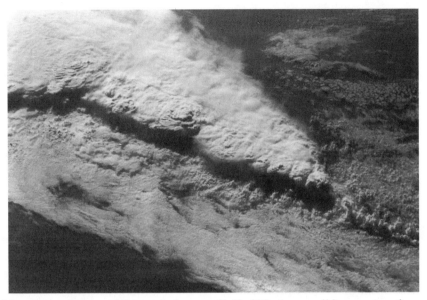

Merging anvils spread downwind over a Gulf of Mexico squall line, as seen from the space shuttle. NASA photo courtesy of Gregory J. Byrne.

Illinois, Indiana, and Ohio. In the tropics, clusters of thunderstorms form regularly in the converging air near the Equator. They look more ragtag than the highly organized MCCs of the plains east of the Rockies, Tibetan Plateau, or Andes, but even the most ordinary-looking tropical MCCs can turn into hurricanes.

Even though midlatitude MCCs don't turn into hurricanes, they can whip up hurricane-strength winds. John Cannon worked the late shift at the National Weather Service radar office in Berne, New York, the night the screen lit up like never before. "The first velocity images were so astounding we had to recheck the radar's calibration system," he later admitted. But the new Doppler dish was getting its signals right in the predawn hours of July 15, 1995. A rare thunderstorm outbreak formed along the edge of hot air that kept overnight temperatures in the 80s in upstate New York. Air in the Adirondacks was so moist that even the clear sky glowed lightly on the radar, as if an invisible fog blanketed the region.

The boundary where the storms formed was several hundred miles distant in Ontario, yet these were no arctic storms. These were full-blooded, 70,000-foot-tall thunderstorms looming over the horizon at 4 a.m., an unusual time for storms to form. They didn't just form, they began to coalesce into a ferocious MCS and raced southwest toward the border at speeds topping 85 m.p.h. Cannon's lightning detection displays went bonkers, recording a strike every second as the line of storms crossed Watertown.

Campers in the mountains awoke to a whistling, a roar, and an earth-shaking crash as gusts accelerated to more than 100 m.p.h., knocking down millions of trees into piles of splintered wood more than a dozen feet high. Four people were killed, and survivors probably thought at first that they had endured a tornado. But this was much bigger, much longer-lived, and much straighter. The high winds covered an area more than 100 miles wide and several hundred miles long. In some areas, winds topped 55 m.p.h. continuously for 20 minutes. After four hours the storms finally died over the Atlantic.

The storms had generated a full-blown derecho, a damaging blast of winds produced by long-lived thunderstorm groups and sustained for hundreds of miles. On average, more than a dozen derechos strike the United States each spring and summer, mostly around an alley of thunderstorms stretching from southern Minnesota to northern Ohio. Gustavus Hinrichs, an Iowa weather expert, first called sustained straight-line thunderstorm winds derechos in 1888 to distinguish them from the curling, twisting tornadoes that do similar damage on a smaller scale. Some derechos can last the better part of a night and blow for nearly 1,000 miles.

Derechos develop under clusters of thunderstorms that form on volatile afternoons with moderately strong, low-level wind shear. Together the storms in an MCS produce a large, uninterrupted pool of cool air under their precipitation. The pool becomes a derecho when it spreads out as damaging winds ahead of the storms. At the same time it continues to scoop warm moist air up to sustain the MCS. Why some MCSs develop derechos while others have more moderate, 25- to 50 m.p.h. gusts ahead of them, still baffles meteorologists. But a quarter of the powerful MCCs do make derechos. Careful analysis of satellite images, radar scans, and balloon soundings shows some of the characteristics of derechos. Some of the derecho-producing convective systems form along a cold front and basically make powerful squall lines. But most derechos strike when the storms organize themselves without much help from a low-pressure system.

A major derecho crossed Nebraska and half of Iowa in just 12 hours on July 8 and 9, 1993. It was a convective system as costly for its rains as for its winds. This pack of thunderstorms leapt up from a convergence of winds north of a weak low-pressure system in eastern Colorado, then sped through a stationary front along the Nebraska-Kansas border. The derecho did not use the front as a crutch; it trampled it, huffing and puffing cold air that temporarily blew down the frontal temperature gradient. The storm drew in the excessively moist surface air hovering over the Plains that summer because of weeks of pounding rains from MCSs. Not only did the Missouri and upper Mississippi rivers flood that year, but the whole state of Iowa was so wet that satellite moisture sensors saw it as a sixth Great Lake.

If the stout cool pool plowing through this moist air is going to make a derecho, the precipitation-laden downdrafts alone are not sufficient fuel. Recent computer models show that the cool pool helps itself out. As in many squall lines, the intense spreading of cold air at the bottom of a derecho system helps stretch the convective towers and strengthen a rear inflow jet. This feedback means the rear inflow jet propels the derecho ahead even faster. So does the northwesterly flow higher up that often drives the MCS into moist and warm air. This exacerbates the violence of the derecho. The faster the MCS, the more violent the crash of the cold pool against warm air ahead. Some derechos bolt ahead of their steering flow like dogs at a fox hunt. A 55-m.p.h. low-level jet streaming from the southwest shoved the 1993 Nebraska derecho from behind. The derecho averaged 45 m.p.h., but also reached top speeds of 70 m.p.h. in Iowa.

The speeding center of the derecho, pushed by the rear inflow jet, can get slightly ahead of the flanking storms. As a result, the Nebraska derecho was a bow-shaped line of storms. On radar displays an arrow was drawn behind the bow: this was the rear inflow jet creating a relatively dry channel through the copious rain from a broad shield of stratus. The center of the bow, at the tip of the arrow, was the strongest part of the derecho. As it swept through Lancaster County, Nebraska, high winds destroyed five mobile homes and damaged dozens more. More than 1,300 trees were uprooted, and two semi trucks were swept off their wheels. More than 200 homes and businesses suffered damage in Omaha. The 90-m.p.h. winds demolished 64 mobile homes in a single neighborhood of Kearney, Nebraska. In Bartlett, Iowa, 85-m.p.h. winds blasted a church off its foundation and blew out windows in cars and buildings in Des Moines. Behind the destructive winds, rains flooded local creeks and rivers that were already swollen by that spring's now infamous weather pattern.

In addition, air swirled around the edges of the derecho, creating volatile tornado-prone zones. Three tornadoes formed near Lincoln on the north end of the intense bow storms. On the south end, in the evening, the derecho storms had no swirl but plenty of effect. The outflow of the derecho restarted a sputtering monster thunderstorm that had been lingering near Hill City, Kansas. Finally, in central Iowa, the derecho choked on air that had already been cooled and stabilized by hours of afternoon convective activity. It was only right that the system stopped there. The earlier thunderstorms had flattened a farmstead with a microburst, pummeled Crawford County with baseball-size hail, and dropped eight inches of rain in only three hours. The water spilled over creek banks so fast that some houses in Dow City flooded in less than 15 minutes. Streets in Chelsea were under seven feet of water in spots.

Fast-paced convective systems causing derechos are able to maximize the interaction between thunderstorms in order to survive. Another type of thunderstorm system survives by interacting in a relatively static environment. These storms conspire with terrain, as happened over the Black Hills of South Dakota in 1972 when a wall of water rushed through Rapid City. It happened again on July 31, 1976, over Big Thompson Canyon, the eastern entrance to Rocky Mountain National Park. About 5,000 flash floods strike the United States each year, but the Big Thompson flood was one of the worst, killing 139 people.

A backdoor cold front (an infrequent, southwestward-moving boundary) marked the advance of cool Canadian air southward from the Dakotas into Nebraska and eastern Colorado that morning. There it met moist air from the Gulf, which it rudely shoved toward the Rocky Mountains. The air lifted along the slopes, and by about 5:30 p.m., the cold front itself rammed into the mountains, bringing with it cool but moist air. A second wave of cool air from the west followed behind it with even cooler, moist air. At this point the situation over the foothills was a loaded gun of potential instability. The layer of cold frontal air was pinned under warmth and moisture pumped in from Arizona storms the previous day, now hanging over the Rockies. All that low-level moisture needed was a little push to 16,000 feet and, filled with heat from condensation, it would rocket upward. This tense standoff didn't last long.

Downdrafts from one thunderstorm, near Denver, pulled the trigger at around 6:30 p.m.: a 40-m.p.h. outflow from this storm rippled northward along the mountains, lifting everything. A wall of dust swept through the area at its edge and behind it temperatures dropped 20 degrees in minutes. Above, new thunderstorms shot skyward to 60,000 feet, forming a 60-mile-long wall of cloud from Boulder to north of Fort Collins.

The Big Thompson storm itself blasted more than 10,000 feet into the tropopause, a highly unusual feat of strength for a storm in that region.

Under the shadows of the building clouds, thousands of campers were enjoying the weekend. At Estes Park, at the foot of the canyon, people celebrated Colorado's centennial. However, none of the fireworks were as spectacular as the rainfall nature provided. A light rain turned into torrents at 8 p.m. More than 10 inches fell in a few square miles over the Big Thompson Canyon in about four hours. At its peak, the storm dropped five inches in only half an hour. Like the Black Hills storm four years before, the thunderstorm was stuck. Low-level winds behind the front jammed it against the Rockies, and extremely weak upper-level winds simply couldn't budge the system. The rain was squeezed into an area only six miles wide. The people in the canyon didn't even see a drop of this highly localized rain in the upper slopes. In the dark, they barely saw the creek begin to rise beside U.S. highway 34, where midsummer waters are usually only about a foot or two deep.

The sides of Big Thompson Canyon became a veritable waterfall. Instead of absorbing the new moisture, the rocky slopes simply passed it down, at speeds approaching 90 m.p.h., toward Big Thompson Creek. In a few minutes, the stream surged with 200 times its usual water. Now more than twice the flow of the Mississippi River at Minneapolis squeezed through the canyon. At 8:35 p.m., State Patrolman Bob Miller radioed his office to report on rockslides. "The whole mountainside is gone.... I'm going to get out of here before I drown." He escaped just as the water rose 19 feet, overflowing the road. A 30-foot crest filled with boulders and tree trunks tore through the canyon at 50 m.p.h., gouging out a 10-mile section of the highway. Another patrolman, Hugh Purdy, was killed in the raging waves. His mud-scoured, water-flattened car was swept eight miles down the canyon and later identified only because someone found Purdy's State Patrol key ring. The officers nonetheless were able to give some people at the motels and campgrounds along the highway a few minutes' warning before the flood. More than 1,000 people wisely climbed the canyon walls and spent the night with frightened rattlesnakes before being rescued by helicopters the next day. In the turmoil below, the muddy flow destroyed 400 homes. The force of the flood dislodged a 200-foot section of steel pipe full of water that weighed more than 500 tons.

The storms over Colorado that day worked up a frenzy with the help of mountains and weak winds that sustained the updrafts. Some storms, however, can take care of themselves. On May 5, 1995, a squall line developed over the Texas Panhandle on the dryline, a common boundary of converging moist and dry air over the Plains. The storms moved with surprising determination southward, toward one of the nation's biggest metropolitan areas, Dallas–Fort Worth. That afternoon 10,000 people gathered in Trinity Park for Mayfest, an annual celebration with food and fun in Fort Worth. Countless others throughout the city were out enjoying other Cinco de Mayo festivities.

At 6 p.m., Mayfest revelers were gearing up for the final concerts in the park when the weather took a disturbing turn. A solitary storm forged ahead of the main line into the warm, moist air flowing in from the Gulf of Mexico. This thunderstorm was the biggest tiger cub in the litter and gobbled up the best food first. Soon it began to spew small hail over northern Tarrant County, not far from Fort Worth. At the National Weather Service office in Fort Worth, chief meteorologist Skip Ely and other forecasters knew this bully storm was a problem when they saw it develop by itself. A severe storm like it had pounded Dallas–Fort Worth airport a few days before with hail as big as softballs. The storm damaged more than 100 airliners, punching holes in wings and tails. More hail fell at a Rangers–Indians game in Arlington, and at the nuclear power plant at Comanche Peak funnel clouds were spotted in high winds. Now, a large, solitary blob on the Doppler radar again threatened the area. The solo storm was a supercell, the ultimate self-perpetuating thunderstorm.

The Texas supercell of May 5, 1995, turned out to be one of the worst in history. Traffic slowed to a halt on Interstate 30 near Arlington as the hail and high winds pummeled the cars. Every car caught in the hail swath had dents; many had smashed windshields. Three teenagers in a pickup saved themselves by crawling into the camper attachment in back just as their truck's windows blew out, shattering glass across the seats. A man driving home from a dealership with his brand-new minivan was lucky enough to park under an overpass when he saw the clouds coming. The rumble of the storm blended with the wailing of thousands of car alarms triggered by the hail. At City Hall in Fort Worth, skylights were smashed. In parking lots, 70-m.p.h. winds bounced cars against cars.

At Trinity Park, the gray sky quickly turned black at 7:15 p.m. As rain turned to hail, people were already running for cover. In only five minutes the hail was as big as softballs and falling at 100 m.p.h. The park's many trees saved hundreds of people, and police broke down the doors of a nearby National Guard armory to shelter others. Two other policemen threw their bodies over a dozen small children to protect them from the hail. Those caught in the deluge suffered welts on their backs. Spiked stones gashed others in the scalp. More than 200 people were injured, and dozens were hospitalized. "Like a rockslide," one survivor said. "Biblical," said another. Another carried an orange-size chunk in his pocket as a memento, the ice seemingly refusing to melt on this spring day.

In Fort Worth residents were lucky the upper winds took the storm to Dallas in less than an hour. In Dallas, the storm halted and merged with the squall line. The hail stopped falling, but the enhanced squall line drenched the Dallas area with heavy rains—some neighborhoods got more than six inches in less than four hours. Roofs collapsed, including that of the city's art museum downtown, and the streets began to flood. One man had parked his car at home, but it floated two blocks away during the storm. Some vehicles ended up stuck on top of others when the waters receded from the parking lot at the Fair Park theater. Catfish from a local pond swam out onto the streets. A teenager rescued one and kept it in a bathtub until the rains were over. When floodwaters covered Samuell Boulevard, Casey Rose waded to the stalled cars and pulled passengers out, and then sheltered them at his business nearby. A dozen people at another flooded intersection waded to safety through an underpass by forming a human chain. But the waters were swift, and in the darkening skies the floods proved deadly. One man rescued a family by towing their car out of a creek bed, where it was stuck in rising waters, but he drowned when his own truck then slipped down the muddy bank into the current. Another person waded out from a stalled car and was sucked into an unseen manhole underfoot.

In all, 16 people died in the storm, most in the floodwaters. For car and home insurers, the toll mounted to $2 billion, a U.S. record for a single thunderstorm. But for Texas weather, it was just another of many deadly supercells. Whether alone or in groups, these long-lived storms are notorious for high winds and large hail. But meteorologists have been tracking and modeling supercells with special diligence over the last few decades for another important reason. They have watched these long-lived thunderstorms spawn the world's most powerful winds—tornadoes.

CHAPTER EIGHT

Monsters of the Prairie

Spencer, South Dakota, is just a one-square-mile island of homes in a vast sea of corn, wheat, and soybean fields. It's not on the interstate, and few people have reason to visit. The young have mostly left this small farming village; so have the stores and schools. On May 30, 1998, only about 320 residents remained—many probably finishing up a quiet dinner—as a whirl of hurricane-strength winds sped in from the northwestern horizon. This outer ring of winds was a mere breeze compared to what followed. At the core of the mile-wide maelstrom was a 200-foot wide vortex with 240-m.p.h. winds.

Face-to-face with the planet's fiercest winds, residents of Spencer had to fend for themselves. It was a moment for life-and-death decisions. One couple hurried to their basement, but the wife paused to look for their dog. Seconds later the twister struck. The kitchen door fell on the husband, pinning him to the ground. After the roof blew away, hail battered his face and body. The house disintegrated, but he survived and so did the dog. The woman, however, was crushed in the debris.

That only six people died in the tornado seemed like a miracle. Only a dozen homes withstood the 100-second onslaught. Not a single business survived. All five churches were demolished. At Security State Bank, only the pyramid-shaped vault rose above the ruins. The wind tore away the walls and roof around it. A blank check from the bank fluttered to the ground 100 miles away, in Minnesota. The first emergency volunteers to arrive brought dogs to sniff the wreckage for survivors. Residents had never needed maps to get around their town before, but they toted them now to help rescuers locate where houses once stood.

Tornadoes don't only prey on people in small, isolated towns like Spencer. But nowhere is a community smaller and more helpless than in the long shadow of a tornado. History is full of these encounters: Woodward, Oklahoma, where 107 or more died in 1947; Saragosa, Texas, where 30 died in 1987, and even little Lake Carey, Pennsylvania, nearly com-

179

pletely demolished, while two people died, a few weeks after Spencer. These towns are emblematic of the sheer power of whirlwinds.

Unbeknownst to residents, however, Spencer did not face the tornado alone. Just a mile outside town, scientists from the University of Oklahoma had parked a small truck full of computers and meteorological equipment. They had driven more than 300 miles to document Spencer's tornado with their Doppler radar. Specially calibrated for fine-scale, close-range scans, the radar probed every movement in the storm, every horrible twist of the tornado.

Professor Josh Wurman and his radar team were among several tornado chasers who made the trip independently. They couldn't have done it without a very good idea about what kinds of thunderstorms make tornadoes. In this case their early-morning hunches about severe weather in the northern Plains nearly didn't work out. Most of the afternoon passed without a major eruption of the volatile atmosphere over South Dakota. In the predawn hours that day the still, humid air shuddered from the effects of thunderstorms in North Dakota. The storms rippled the air southward with waves, spreading out a bowlike leading edge that triggered a new crop of convective towers bursting through a cap of warm air. These new storms northwest of Spencer were relatively benign until midafternoon. For a while nothing much happened.

Then a cold front swept in from the west. The western end of the line of formerly complacent storms recoiled northward. Shocked by the wedge of new air, the towers shot skyward. Tornado chasers could see on satellite images that a couple of storms in particular punched into the stratosphere with special persistence. Their anvils dominated the sky, flashing defiant V-shaped cirrus outflow for the benefit of satellite sensors. The V's meant that the storms generated remarkably sturdy updrafts that were disrupting fast, spreading upper-level winds that pushed the cold front in from the west.

One V storm produced golf-ball-size hail at 6:30 p.m. as it moved southeastward with its cohorts. Another approached Spencer, and Wurman's chase group had it pegged pretty much from the start. Just after 8 p.m., near Fulton, South Dakota, they aimed their radar at the thunderstorm cloud as it spun up the tornado (one of five that day). As the clouds moved east, so did the chasers. A half-hour later, the storm merged with fast-rising clouds next to it, forming a monstrous tornado-maker over Spencer. The tornado churned the landscape below for 21 miles, and all the while the V-shaped anvil over the merging storms grew even stronger,

The V-shaped anvil of a powerful thunderstorm spreads while the tower of convection overshoots it by nearly 10,000 feet into the inhospitable stability of the stratosphere. These anvils are often a sign of tornadic storms. NASA photo courtesy of Gregory J. Byrne.

dominating the small line of convection as it gathered momentum. The strengthening storms formed an MCS that swept the Midwest and Great Lakes that night, eventually reaching as far as New England. Near Muskegon and Grand Rapids, Michigan, downburst winds reached 130 m.p.h.

Meteorologists can now document the development of a tornadic storm as never before. Their new abilities to forecast, simulate, and observe tornadoes mark a new era. Before the age of storm chasing and computer modeling, scientists couldn't see tornadoes for themselves. All they could do was marvel at the stories survivors told afterward. The evidence pointed at incredible forces at work. Now scientists are edging closer to explaining the once nearly inexplicable power of tornadoes.

In 1953, one of the Weather Bureau's experts on tornadoes, Snowden Flora, wrote a book recounting what meteorologists knew up to then. Flora extensively quoted a prominent airline weather researcher, Henry Harrison, who had concluded that tornado winds might even be supersonic—

more than 700 m.p.h. At the time it was the only plausible way to explain how tornadoes could make delicate blades of grass fly like arrows into tree bark. Or how a Scottsbluff, Nebraska, tornado shot a bean into an egg, making a clean hole without a crack. A study of the Woodward, Oklahoma, disaster in 1947 indicated a similarly overblown wind speed of 454 m.p.h. on the right side of the tornado.

This was also the degree of force meteorologists thought necessary to explain the famous May 27, 1931, tornado that collided with a train east of Moorhead, Minnesota. The Empire Builder steamed eastward at 60 m.p.h. when the twister broadsided it. Five 70-ton coach cars jumped out of their couplings and off the rails. One car, with 117 people aboard, flew into a ditch, landing 80 feet from the track. Miraculously, only one person died. A dozen years before, a tornado destroyed half of Fergus Falls, Minnesota, and went on to throw seven of the coach cars of the Oriental Limited off the track. In a final cleaning frenzy, the tornado wrenched the rails off their bed as well.

The average tornado is almost a quarter of a mile wide and travels at a little better than 30 m.p.h. But tornadoes can last for hours in extreme cases. One traveled 293 miles through Illinois and Indiana on May 26, 1917. The worst in American history, the Tri-State Tornado, ripped up a 219-mile path from Missouri to Indiana on March 18, 1925. It caught most of its 689 victims by complete surprise. Some people thought the funnel was a curious fog under the storm cloud and took no heed at first. Others, in the railway towns along the tornado's route, thought it was a train belching black smoke. Unfortunately this dark cloud roared into towns at top speeds of 73 m.p.h. and bore dirt, books, stones, planks, chickens, glass, and other lethal missiles. After the tornado destroyed 100 blocks of buildings in Murphysboro, Illinois, the black skies grew even darker as a wind-whipped fire incinerated another 70 blocks. In Parrish, Illinois, whole houses popped off their foundations. The twister stripped trees bare of their leaves and impaled victims, mostly women and children, in the branches. After the storm, local miners emerged from underground to a stunning scene of devastation; to many of the men, veterans of war-ravaged Europe, the haunted, scorched landscape looked dishearteningly familiar.

For all the wrath of a tornado, its interior may be remarkably calm. Flora and scientists of his day feared this calm, an intensely rarefied core where atmospheric pressure can drop about 10 percent. In Flora's day, scientists mistakenly believed the sudden loss in exterior pressure explained why the walls of a house tended to fall outward in a tornado—the structure must explode. The theory seemed consistent with other findings, like the tornado that sucked the water out of a 39-foot-deep well in the 1800s.

Because it is calm, some people have lived to tell us something about the eye of the fury. Will Keller, a farmer near Greensburg, Kansas, survived a June 1928 tornado. As the thunderstorm approached, Keller saw a "greenish black base" under an arcing shelf of cloud and quickly hurried his family into their shelter. Before taking cover himself, he couldn't resist one more look. Some people never live to face temptation again. But Keller was fortunate, and as a result so were meteorologists.

Three funnels approached him, "like great ropes dangling from the parent cloud." Keller was mesmerized as the center funnel moved overhead. "Everything was still as death," he recalled. The air smelled of gas, and Keller had trouble breathing as the tornado overtook him. The noise from the end of the funnel was a "screaming, hissing sound." (Others caught nearby have recalled the sound of a million bees, or a roar of many freight trains or jet planes.) Keller looked up "right into the heart of the tornado," a round opening less than 100 feet across. The vortex tube appeared to be a half-mile tall, surrounded by a rotating wall of clouds and crisscrossed by lightning, which illuminated it from within. Smaller tornadoes formed and briefly moved around the rim before disappearing. Inside, a bright cloud bobbed up and down.

Keller's account is remarkably similar to another from Roy Hall, whose survival story appeared in *Weatherwise* magazine in 1951. Hall and his family were at home in McKinney, Texas, on a warm, humid spring day when distant thunderstorms, "black as ink," crept up on them from the west. At the same time low scud clouds swept into the darkening storms from the south. Hall, an avid amateur meteorologist, and his wife were sitting on the porch watching the conflicting winds. The convergence made them nervous. "What a terrible cloud!" Hall's wife exclaimed when she saw the storm, a few miles off. She hurried their children inside, while Hall watched the scud circle under the storm. Below the thundercloud was a "curtain of dark, green rain." A distant roar began to crescendo. In Texas Hall had endured plenty of hail; he knew this rumble meant something else. Never before had he felt the air beat against his face.

Soon lightning flashed about the house and baseball-size hail banged against the roof. Some of the ice crashed through the shingles. Hall ducked inside just before the wind-driven rain slammed against the windows. The west wall blew in a few inches and began to flap like paper under the pressure. Hall yelled to his family to hide under a bed in the back, but, like Keller, he decided to take one last look. In the darkness he saw his neighbor's trees bow to the ground, blown eastward where moments before the wind had been from the south.

All of a sudden the lightning and din stopped—"exactly as if hands had been placed over my ears," Hall wrote. But the house still shook as the room suddenly brightened. Hall thought a fire had broken out, but then realized everything was bathed in a "bluish tinge." Papers and magazines hovered above him and the window curtains pressed flat to the ceiling. Instantly, Hall was thrown 10 feet across the room onto his back as the house lurched off its foundation and flew whole into the trees in the yard. Everything loose in the air simply disappeared; part of the roof was gone, too: "The side of the room came in as if driven by one mighty blow of a gigantic sledgehammer."

A bluish glow illuminated the inside wall of the tornado, its rim maybe 10 feet thick and just 20 feet off the ground. The funnel was smooth and a blur with high-speed rotation. Lying on his back, Hall could see about 1,000 feet up into the opening, where a "bright cloud hovered inside and shimmered like a fluorescent light." High in the tornado, the walls expanded the diameter far beyond the 150 yards enclosed at the bottom. The "huge pipe swayed" back and forth overhead, its sides rippled with rings. Then the back edge of the tornado approached, and "long, vaporous, pale-blue streamers" about 20 feet long leapt upward from the corners of the house. Suddenly, the light was gone and Hall was plunged into the blackness of the storm again. Debris pounded the house, and he could hear his family talking excitedly. In a few minutes daylight returned. The Halls survived, though nearly 100 people died in the storm elsewhere.

Stories like Hall's and Keller's led more scientists to suspect lightning played a role in tornado strength. In his study of tornadoes in the 1940s, meteorologist Henry Harrison claimed that tornadic thunderstorms had unusually strong and frequent lightning. Since lightning creates static at radio frequencies, NOAA researchers in the early 1970s tried in vain to isolate the special pattern of electromagnetic pulses in tornadic thunderstorms. They found that the frequency of pulses might be linked to the strength of the storms. But this line of research didn't go very far. Even television seemed to hold promise for tornado warnings at the time. In 1967 an Iowa inventor named Newton Weller showed that a blackened TV screen would glow brightly on Channel 2 during the passage of a tornado. But the method turned out to be less than a sure thing: some storms caused the glow without a tornado and other storms caused tornadoes without the glow.

Strictly speaking, meteorologists turned to electricity because they couldn't explain the unimaginable winds. They knew the tight rotation of a tornado formed when a broader circulation of air collapsed about its axis. The air conserves its angular momentum, driving it faster and faster. A figure skater does the same thing by drawing in his or her arms to accelerate a spin. The smaller the radius, the faster the skater twists. But the tornado's winds exceeded the speeds meteorologists calculated if a large storm's winds drew inward and conserved angular momentum (minus the effect of friction). Some other force seemed to be working in a twister. In the 1830s, the chemist Robert Hare and others had insisted that electricity drove tornadoes. In the 1960s Bernard Vonnegut—already famous for his cloud-seeding work—insisted that cloud electrification could spin up tornadoes. The logic seemed plain enough: tornadoes come from thunderstorms, which are defined by electricity. No other unaccounted force seemed large enough to do the job. Vonnegut suggested that the tornado might sheathe a nearly continuous discharge of electricity. This might also explain occasional glowing discharges and scorched plants and soil associated with tornadoes. In 1967 Marx Brook added more fuel to Vonnegut's electrification fire. Brook, a lightning expert who often worked with Vonnegut in New Mexico, published data showing that a tornado deflected the electromagnetic field around it.

Stirling Colgate, an astrophysicist at Los Alamos Labs, wanted to test the lightning theory in tornadoes without actually walking inside. In the early 1980s, he chased tornadoes in an airplane loaded with air-to-air missiles. Colgate hoped the rockets would yield badly needed electrical field measurements inside a tornado. Though he was able to maneuver close to a few tornadoes, only a few missiles hit their target. Most of them veered off target, soaked by rainfall or battered by wind shear. FAA rules, which limited Colgate to two-pound missiles in civilian airspace, severely crimped the project. But flying a little Cessna prop plane near major thunderstorms also proved treacherous. In 1982, Colgate had to make an emergency landing in a field after the plane got knocked around in the low-level winds racing 60 m.p.h. into the underbelly of a thunderstorm.

Very few researchers think lightning explains tornadoes. They point

A tornado—probably packing winds stronger than 200 m.p.h.—tore through the historic district of Petersburg, Virginia, before crossing the Appomattox River and killing three people in this store. Photo by Mark Gormus/Richmond (VA) Newspapers.

out that the energy of a lightning flash is concentrated in a tiny channel, hardly able to affect wind motion on the scale of a tornado. Furthermore, lightning doesn't consistently increase during tornadoes, nor does the ground often get scorched. In fact, many times lightning stops striking the ground frequently while a thunderstorm spawns a tornado. Of course, it is nearly impossible to completely disprove Vonnegut's suggestions. But more likely explanations for tornadoes supersede lightning. In part this is because wild speculation about the strength of tornadic winds ended. Ted Fujita's cool and thorough studies of tornadoes were an important factor in the new sanity in tornado theory.

Fujita was Mr. Tornado in his heyday from his arrival in the United States until semiretirement in the early 1990s. Nearly every significant discovery in the field has his fingerprints on it. Like a detective, the University of Chicago researcher didn't need to see the crime, just the crime scene. In fact, he didn't see his first tornado until he flew near one while studying microbursts in 1982. Instead, he tracked the footprints of tornadoes in the debris, interviewed the witnesses, performed lab tests, and found the culprits. He followed a long line of damage analysts dating back to the bickering scientific gumshoes who let the Brunswick Spout of 1835 get away. James Espy and William Redfield may have made a mess of that survey, but two years later, Elias Loomis carefully analyzed a Stow, Ohio, tornado and showed that damage analysis could yield reliable results. One noteworthy find was the grisly evidence of the tornado's power. The bodies of turkeys, geese, and hens littered the farms, and "several of the fowls were picked almost clean of their feathers, as if it had been done carefully by hand."

Loomis realized this evidence offered scientists an opportunity to confirm what sorts of winds a tornado could muster. So in 1842 the Western Reserve professor performed an early meteorological simulation. He killed a chicken, quickly stuffed it into a small cannon packed with powder, and blasted it into the air. The feathers indeed separated cleanly from the bird and floated away 20 or 30 feet in the air. Unfortunately, "the body was torn into small fragments, only a part of which could be found." The flightless bird had just made its first, and last, flight at 341 m.p.h. A tornado, Loomis figured, wouldn't need to blow quite so hard. True to form, it was the ever-curious Vonnegut, in a *Weatherwise* article of 1975, who half-seriously proposed redoing the experiment under modern conditions. But a review of the biological literature, he said, showed that in

The mysteries of tornadic winds continue to puzzle researchers. A 1984 killer twister wrenched this half-ton I-beam from the roof of a garage and carried it a quarter-mile before sinking it 8 feet into this rain-soaked lawn in Bennettsville, South Carolina. Photo by Jim Campbell/National Weather Service.

fact chickens lose their feathers under extreme stress, a far more likely explanation. No need to ruffle any more feathers.

Fujita brought his considerable map-making skills to tornado surveys. His colorful charts recall the artistic care Jacob Bjerknes and Tor Bergeron lavished on their charts. Fujita showed details like the flight patterns of diving boards and I-beams and the paths of cars as they tumbled through cornfields. He also used aerial photography for his work. In the Palm Sunday tornado outbreak of 1965, 36 tornadoes struck six states, killing 253 people. The next day Fujita took off in a small Cessna to photograph as much of the damage as he could before survivors began to clean up. He covered 7,500 miles in four days and showed that the long-tracked tornadoes were actually families of shorter-lived tornadoes. When one tornado lifted, the thunderstorm would spawn another along the same track. The discovery made clear that many of the long-track tornadoes of the past, like the Tri-State, might very well have been families.

Fujita also studied a famous series of photos of one of the Palm Sunday tornadoes near Elkhart, Indiana. The funnel appeared to split into twins, but Fujita's detailed damage analysis showed instead that one tornado formed independently of the other, then the two swung around one another in a country dance. The Hesston and Goessel tornadoes that killed two in Kansas in March 1990 were similar binary tornadoes. At the end of its 60-mile solo show, the Hesston tornado briefly sped up to over 72 m.p.h. The Goessel funnel slowed temporarily to compensate, as the two began to circle a common center. It was as if the Goessel twister were a pairs figure skater turning gracefully to catch his leaping partner, then swinging her to his opposite side. The older tornado gracefully dissipated after the four minute interlude.

Fujita and his University of Chicago crew surveyed more than 300 tornadoes over his career. Nearly half struck in a massive outbreak in 18 hours on April 3–4, 1974. In all, 148 tornadoes tracked 2,584 miles east of the Mississippi River, from Mississippi and Alabama north to Michigan and east to New York and Virginia. It was a day for the bizarre as three parallel lines of squalls more than 11 miles high raced at about 50 m.p.h. across the landscape. One of the worst tornadoes covered 51 miles in Alabama, crossing a lake and thrashing a mobile home park with perhaps 260-m.p.h. winds. A man injured in that storm sat at a nearby church to await an ambulance. Within a half an hour, a second tornado demolished the church and killed him. The strongest tornado of the outbreak destroyed or damaged almost half the houses in Xenia, Ohio, a town of 25,000. Thirty-four people died, and 1,150 were injured. Had the late-afternoon tornado occurred an hour earlier, Xenia would have suffered far worse. Six schools were damaged, three of them wrecked completely. A high school drama troupe dashed into a nearby hallway just before the auditorium roof collapsed behind them. The tornado tossed three school buses onto the stage, for good measure.

In all, 315 people died and nearly 6,000 were injured, but plenty of people made remarkable escapes. Near Branchville, Indiana, a school bus rolled 400 feet off the road, killing the driver and his wife, but the driver of another bus there saw the tornado coming and ordered the children to evacuate and lie in a ditch. The bus blew over them and no one was seriously hurt. A man driving home from work pulled onto his street as the tornado bore down on his house. The car somersaulted twice and landed in the neighbor's yard. The man was badly cut by glass but hurried into the remains of his home to find his family had survived, in the basement beneath the rubble.

Fujita flew more than 10,000 miles on the following days, photographing as much as he could. The surveys yielded a "goldmine" of deadly data. As in the Palm Sunday outbreak, most of the tornadoes occurred in families, especially the killer tornadoes. Individual tornadoes had a resiliency the Terminator would envy. One twister climbed a 3,300-foot peak in Georgia and then swept into the valley below. Another jumped a 200-foot cliff in Alabama and kept right on going. In 1987, Fujita's staff surveyed a similarly amazing tornado that crossed the Continental Divide at a pass more than 10,000 feet high—on a 24-mile run through the Teton Wilderness and Yellowstone National Park. The swath through the forests averaged a mile and a half wide—15,000 acres in all. The tornado ripped the bark off the trees it leveled when winds peaked at 200 m.p.h.

While surveying the record 1974 outbreak, Fujita photographed a peculiar, starburst pattern in one forest. The trees all lay pointing outward from one spot, a seemingly impossible divergence for tornado damage. When evidence of powerful downdrafts emerged in his investigation of the crash at Kennedy airport in 1975, the photo proved invaluable in making the case for microbursts. In later surveys, Fujita recognized evidence of damaging downdrafts and explained several suspected tornado swaths as the work of microbursts instead. By separating damage due to microbursts from that due to tornadoes, Fujita's surveys revealed that tornadoes may zig and zag suddenly owing to violent thunderstorm outflows. A tornado that crossed from Connecticut to Massachusetts in 1979 repeatedly veered sharply to the left after passing by Bradley International Airport. Fujita's staff showed that seven microbursts from the parent storm had dropped in the rainy downdraft beside the funnel and knocked it off its path. Each microburst temporarily dislodged the neatly converging wind pattern associated with the tornado.

To refine tornado statistics, Fujita and Allen Pearson, then director of the National Weather Service's severe-storm-forecasting group, developed a standardized scale for judging tornado damage. The accumulated experience of tornado path flights and engineering analysis led Fujita to attribute damage to winds far weaker than those Flora and Harrison had imagined two decades earlier. On the Fujita scale, the weakest tornadoes, F0, break branches and knock down trees with shallow roots with 40- to 72-m.p.h. winds. An F1, with 73- to 112-m.p.h. winds, can shove a car off the road or a mobile home off its foundation. At F2 (113- to 157-m.p.h.), a tornado can uproot a large tree or tear a roof off a wood-frame house.

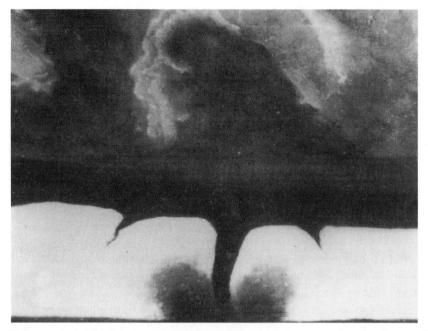

This snapshot—the first known photograph of a tornado—was taken southwest of Howard, South Dakota, on August 28, 1884, not long after Sgt. John Finley's pioneering Signal Service forecasts. NOAA photo.

This scale reveals what engineers (most notably from Texas Tech) found from numerous damage surveys: houses blow apart because of the wind flow into and around them, not because air "trapped" inside pushes out against the low-pressure core of the tornado. The pressure inside most houses equalizes with the outside pressure quickly enough because no house is snugly sealed, even with the windows shut. In fact, you are probably safest with the windows shut. If wind or flying debris smashes through the wall, the air rushes into the house and pushes against the back wall. This pressurizes the interior and the roof might lift just enough to allow the back wall to collapse. Everything tumbles after that. Once the roof blows off, the walls are weakened and collapse quickly. Wide eaves also help F2 winds pry up a roof if it isn't bolted down securely. In Xenia, Ohio, in 1974, all the houses on one side of a street collapsed while houses facing them showed less critical damage. The reason, engineers concluded, was that strong winds bashed in the garage doors—a common weak point—along one row of houses. Once the wind entered the houses they collapsed. Across the street, the same winds blew away from the garage doors, so the roofs remained attached.

Even stronger winds, filled with large debris, can cause far worse damage than the Fujita scale indicates. The scale is calibrated for damage caused by winds, not missiles. When one weak house in a neighborhood fails, well-built houses downwind from it may fail as well, even in a relatively weak tornado. Tornado trackers see damage from large airborne debris in an F3 (up to 206 m.p.h.), which can overturn a train, uproot most trees in a forest, and throw a large car into the air. At F4, 207- to 260-m.p.h. winds flatten a well-constructed home, even pick up a whole house if it is poorly connected to its foundation. The strongest tornadoes, F5, have 261- to 318-m.p.h. winds and carry houses significant distances, hurl a car like a missile for more than 100 yards, and debark trees. "Incredible phenomena occur" during the F5.

Fujita meant his scale to serve as a bridge to a time when technology enabled meteorologists to measure tornadic winds more satisfactorily than by proxy with damage surveys. Radar evidence such as that gathered at Spencer, South Dakota, may eventually provide verifiable wind measurements at the surface exceeding 300 m.p.h. Until then, some scientists have backed down a bit from even Fujita's conservative estimates, noting that no photogrammetric study of extreme tornadoes has verified winds as high as 250 m.p.h. Others, citing the developing resolution of radar readings, maintain that significantly higher winds are likely to be clocked in funnels eventually.

No matter how accurate it really is, the Fujita scale itself is difficult to apply because "well-constructed" depends on local building codes. For instance, the Jarrell, Texas, tornado of 1997 that killed 27 and wiped houses clean off their slabs earned an F5 rating from the National Weather Service. The demolition could not have been more complete. But further engineering analyses showed than an F3 could have achieved the same destruction given the construction techniques used. Local standards were less stringent than those in most hurricane-prone areas. Such discrepancies suggest damage-based wind estimates of the strongest tornadoes may be off by as much as 30 percent.

No tornado damage has ever indicated winds above 318 m.p.h., and in fact most damage suggests winds of about 100 m.p.h. In any case, few would question why Fujita labeled the F6 category "unthinkable." It is unthinkable enough that less than 200-m.p.h. winds can tear off your garage door and then flip cars over inside, as happened once in Texas. Or that a tornado in Louisiana wrenched 750-pound steel I-beams off an elementary school roof and hurled one of them nearly a quarter of a mile. The beams landed so forcefully that they embedded themselves in the rain-soaked ground. One stood tilted 23 degrees and wouldn't budge an inch when Fujita and his colleagues hung off the end.

On April 8, 1998, residents of Birmingham, Alabama, found out why F5 is as high as the ratings go. An F5 there killed 34 people and injured 221 others as it churned 20 miles through the Oak Grove, Concord, and Rock Creek neighborhoods before it literally "carpet bombed" a stretch of Pratt City and Sandusky. *Birmingham News* reporter Russell Hubbard described one 15-block area in the suburbs as a "shattered lumberyard." A local policeman who had been patrolling the neighborhoods for more than 23 years couldn't find his way through the wreckage. The few twisted street signs remaining didn't help much.

At Open Door Church near Ensley, 67 people were concluding an evening meeting when they heard the tornado warning on a radio. For 20 minutes they crouched on the floor and prayed. Then they heard the roar of a freight train, loud hissing, and crackling as the walls split open. The church disintegrated around them, but fortunately only a few people suffered injuries. In another neighborhood, a boy died from injuries sustained when the tornado heaved him 100 yards. In Concord, a woman clutched her baby girl as they hid under a mattress inside. The wind lifted them both and snatched the girl from her mother's arms. They landed, critically injured, in a grove of trees several houses away. A double-wide mobile home flew a similar distance, smashing open when it landed in a ravine. Luckily, the owner was not home.

The tornado's swath through the suburbs looked like "a river" of debris, one house indistinguishable from the next, and stopped just three miles short of downtown Birmingham. Piles of bricks and splinters were basically all that remained of more than 1,000 homes. Whole roofs perched atop debarked trees foliated with clothing and sheets, not leaves. Here and there the walls of an interior bathroom stood among the rubble—the pipe-reinforced walls forming a small sanctuary for a whole family. One man spoke for many survivors when he told the *Birmingham News* it was the most frightening night of his life, admitting, "I've never been so close to my carpet."

The Birmingham tornado was relatively easy to classify. Sometimes tornadoes pass over empty fields, making classifications above F3 very difficult to confirm. In any case, only about 5 percent of the roughly 900–1,000 tornadoes a year in the United States are that violent. And far less than 0.5 percent of all tornadoes are F5.

In 1971 Fujita figured out one of the secrets of the high winds in the worst tornadoes. Many of the F5s and other devastating twisters are actually a collection of subvortices orbiting the tornado core. During the Palm Sunday outbreak of 1965, Fujita pulled out a telephoto lens over Kokomo, Indiana, and took pictures from the air of cycloidal marks in a cornfield. The thin, overlapping circles were hundreds of feet wide. About a decade earlier meteorologists had first noticed such marks from the air and assumed the tornadoes had scratched the landscape by dragging debris, but Fujita's pictures suggested instead that the tornado somehow had piled vegetation into neat looping rows. In a survey of a tornado that struck Barrington, Illinois, in 1967, Fujita again saw such cycloidal rows of corn leaves from the air. Later he visited the site and saw that the marks indeed were not scratches but debris piled about a foot high.

The cycloidal rows in cornfields suggested a very small vortex at work. Fujita called the small vortices within tornadoes "suction spots": Tornado winds blow debris down; the winds around them converge on a scale that can exceed a half mile. But in the smaller vortices, the convergence area is so small—10s of feet—that the vortex seems to suck debris together. Since the suction spot moves around the edge of the greater tornado, its piles create overlapping loops. When a small vortex first spins up along the wall of the tornado, it may leave an eyelike spot of damage—a 20-foot-wide circle where everything converges radially toward the center.

The 1974 outbreak helped prove Fujita's case for suction vortices. Fujita rated six of the tornadoes in the outbreak as F5 (the nation saw only three F5s throughout the 1980s). With home movies of the Xenia, Ohio, tornado, Fujita studied the motion of dust and debris. The tornado was often transparent—the funnel didn't stretch all the way from ground to cloud. Without a thick sheath of condensation, the internal circulation was rather clear, and Fujita could track several smaller dust whirls inside. One of them, only about 35 feet across, spun at 100 m.p.h. while rotating around the larger vortex. When the suction spot moved west to east on the south side of the tornado, three motions coincided in one direction: the motion of the twister as a whole, the rotation of the suction spot around the twister,

Vulcan, Alberta, under the shadow of a tornado in July 1927. The northern prairies of Canada can be more dangerous than Kansas when the jet stream moves poleward with summer. NOAA photo.

and the spin of the suction spot. Repeatedly, in these two second bursts, the tornado created swaths of incredible damage less than 100 yards long and 20 yards wide.

Other tornado photos revealed the suction vortices more clearly. The powerful Wichita Falls, Texas, tornado had them, but the damage surveyors could not find corresponding cycloidal damage paths. Finally, in 1991, Fujita's assistant Duane Stiegler documented a swirling damage pattern left by the Goessel, Kansas, killer tornado. At exactly the same spot, a storm chaser photographed the monstrously dark cloud. Each suction vortex was a pleat in the dense wall.

Meteorologists found they could model the multiple vortex tornadoes in laboratories. Scientists had been making miniature funnels since at least the late 1700s, when water, not air, filled many of the vortex chambers. Some scientists created low pressure by draining water through the bottom of the chamber; others swirled water with rigid wires. Johannes Letzmann, a pioneering tornado researcher from Estonia, drew water upward along a central axis. Working in Graz, Austria, in the 1930s, he turned to a more sophisticated model with water heated from below. In this chamber Letzmann made subvortices rotate around the main "tornado." From witness reports, Letzmann knew that tornadoes had similar complex structures. Unfortunately, Letzmann's research, including his many detailed tornado surveys, languished in obscurity. In the 1960s and '70s Fujita rediscovered and explained many of Letzmann's finds—the miniwhirls, anticyclonic tornadoes, interactions between tornadoes and downdrafts. At about the same time Letzmann's career ended quietly, in the 1950s, Neil Ward, a self-taught scientist with the Weather Bureau, began to revive tornado modeling as a science.

Ward started out as a weather observer for the government and gradually worked his way up to forecasting for the Weather Bureau in Texas. But his real love was tornadoes, and he later joined the Weather Bureau's severe-storms research group (formed in 1964). Ward built a "tornado simulator" in his garage in 1956, using a fan to draw air through a chamber. At the top was a honeycomblike baffle to smooth the flow; the air entered above a rotating wire mesh floor, then passed through a hole into the chamber. In 1972, a year after Fujita formally proposed his suction vortex theory, Ward published results of his modeling, showing the multiple vortices in his laboratory chamber.

Subsequent tornado chambers at Purdue University and the University of Oklahoma used Ward's basic design, generating updrafts about one or two feet in diameter. In order to measure the flow without disturbing it, the scientists eventually resorted to lasers that acted like radar guns. For truly unobtrusive models, however, scientists have turned more and more to computer-based numerical simulations. Lab models and numerical computer simulations show that inflow winds for the strongest tornadoes may exceed 120 m.p.h. They also show that inside the upward-spiraling circulation of the twister wall, vertical speeds can be as much as 175 m.p.h. But in the calm eye of the tornado the extremely low pressure actually sucks air from the parent storm downward. The tornado models—both in glass chambers and on computers—also prove the importance of surface friction: make the ground slippery, and the tornado turns slower.

Using tornado models, meteorologists figured out how real storms might develop multiple vortex tornadoes. The scientists tinkered with the funnel configuration by changing the "swirl parameter," a ratio of the spin of the ambient flow (about the tornado's axis) to the amount of inflowing air. At first, when convergence dominates spin, the vortex approximates a rotating cylinder. Flow stagnates slightly at the bottom, where it turns upward. Crank up the swirl, and spin dominates convergence: the funnel "breaks down." The stagnation at the bottom disappears. The ground flow splashes together and outward. The funnel expands to accommodate a flow sucked down from above. Somewhere perhaps midway down the funnel, some of this sinking air gets swept upward, creating a secondary source of rising, spinning air. Crank up the swirl even more, and the vortex disintegrates into miniwhirls—suction spots. Storm chasers have documented that some real tornadoes widen and perhaps repeat this classic breakdown sequence as they develop suction spots.

Vortex breakdown helps explain the configuration of tornadoes, but not the incredible wind speeds. They remain an open question in meteorology. Brian Fiedler of the University of Oklahoma suggested that twisters break the "thermodynamic speed limit." In other words, heat energy drives pressure gradients, which in turn drive winds. Somehow, though, tornadoes spin faster than these processes allow. The vortex breakdown simulations show that the highest winds are at the juncture where inflowing winds turn upward into the vortex. Converging surface winds might augment low pressure simply by "overshooting" the juncture. The overshooting flow behaves like a strong wind that sweeps past a wall; on the leeward side, a pocket of low pressure forces some of the air to curl back.

Neil Ward's tornado chamber helped spawn much of this theoretical work, as did his observations. Unlike Fujita, who didn't see his first tornado until 1982, Ward represented a new breed of storm scientist. His family knew that their drives across the Texas countryside could turn into impromptu chases down bumpy, dusty roads in pursuit of dust devils. Once, Ward spotted a dust devil, stopped the car, and ran after it across a field. To his chagrin, the devil turned and headed over the car instead. Soon Ward was pioneering a whole new style of storm studies, taking the car on trips dedicated to chasing storms.

On May 4, 1961, Ward arranged to ride with an Oklahoma Highway Patrolman to hunt severe weather. They caught a massive tornado three miles northwest of Geary, Oklahoma. Using the police radio, Ward was able to keep in touch with the meteorologists in Oklahoma City; together they could compare the storm's appearance to the echoes on the radar scope. It was the first scientific tornado chase in history.

Ward certainly wasn't the first scientist to indulge a passion for experiencing the weather. Benjamin Franklin probably launched his kite into a thunderstorm with similar motivations. One day in August 1775, Franklin was riding a horse on a dusty trail in rural Maryland when the afternoon heat stirred up a whirl of dust ahead of him. The bottom of the dust devil was "not bigger than a common barrel," but at its top, 50 feet high, it flared out to 20 or 30 feet wide. The dust devil passed right by Franklin. The first meteorologist wasn't ready to let it go. He quickly spun his horse around, charged in hot pursuit, and then rode alongside the dust devil. Franklin called the world's first storm chase a "real treat."

In the late 1950s, while Ward was driving around Texas watching thunderstorms, David Hoadley began driving around Kansas and Oklahoma taking pictures of storms. Hoadley caught storm fever when a damaging storm ripped through Bismarck, North Dakota. The next day the youngster went out to see the damage and was mightily impressed. Even though Hoadley moved to the Washington, D.C., area, he continued to spend his vacations in the Midwest, where the wide treeless vistas, powerful storms, and empty roads make tornado chasing most effective. In 1965, near Dodge City, Kansas, Hoadley finally had the big thrill of witnessing a tornado.

Eventually, Hoadley started a small storm-chasing magazine, called *Storm Track*, now edited by chaser Tim Marshall. Hundreds of people now subscribe, and the rising popularity of chasing increases each year the chance of injury or death, especially since many of these people end up in the same place at the same time, looking for the same narrow road in case they need to escape in a hurry. State troopers worry, too, that chasers speed to catch storms in treacherous weather conditions, or that chasers block roads needed for emergency vehicles. But most experienced chasers stay away from dirt roads, which turn to mud in a downpour. They refuse to drive into the rainy backside of a thunderstorm, knowing full well they might stumble into a tornado raging in the "bear's cage." Many also help authorities by radioing ahead when they spot troublesome weather. The dangers of chasing so far have mostly been close calls with large hail and lightning.

In one editorial, Hoadley explained why chasing is not really about either danger or nabbing the "big one": a chaser adores "the sheer, raw experience of confronting an elemental force of nature—uncontrolled and unpredictable—which is at once awesome, magnificent, dangerous and picturesque ... eternal—but ephemeral." Watching the storm work its magic, yet understanding how it came about, a chaser can almost participate, "as if—by force of will—he could detach himself from earth and ride the wind up into the storm's core."

But storm chasing has become simultaneously one of the ultimate fin-de-siècle sports for some people and a serious science for others. The reason has less to do with the eternal attractions of transcendence than with the sudden emergence of technical skill and technological feasibility. Marshall points out that without modern communications—radio, TV, faxable weather maps—only the very committed chasers like Hoadley and Ward could take to the roads. And, of course, those roads, the vastly improved interstate system laid out in the Eisenhower years, made it possible for people in Texas to chase storms in, say, Nebraska, and still make it to work the next day.

Storm chasing is also a milepost in the science of severe thunderstorms. Chasers live for the challenge of forecasting tornadoes. It is half art and half science. The best chasers keep up with the latest research—and contribute to it—compelled by the thrill of an accurate prediction. Ward and Hoadley began chasing almost exactly at the time when meteorology, not just highways, made reliable severe-thunderstorm predictions possible. This scientific advance began in the late 19th century. In 1881, General William B. Hazen, the new chief of the Army Signal Service, committed the nation's young weather service to meteorological research. Hazen hired seven civilian scientists and in 1882 organized them into a research branch under 28-year-old Sergeant John P. Finley. No one knows exactly when Finley's interest in tornadoes began. The graduate of State Normal College (now Eastern Michigan University) joined the Signal Service in 1877. As head of the new weather research wing, Finley pushed to complete his monumental study of American tornadoes—an analysis of more than 600 storms dating back to 1794. He completed the study in 1882 but drove himself to nervous exhaustion in the process. The Signal Service refused to publish the study on grounds that it was filled with typographical errors. It was the first salvo in a running battle Finley would wage with the Service. While he was in the hospital, the doctor told General Hazen, "He is doing too much brain work.… Should he continue to do as much as at present, the consequences will be of very grave character."

Finley was a portly 300-pounder who stood 6 foot 3. He may not have been in the best of shape, but he wasn't used to backing down. Two years later, he cleaned up the report enough to have it published. A summary appeared in the prestigious journal *Science*, and another version won a prize from the *American Meteorological Journal*. Meanwhile, Finley recruited a special network of volunteer weather observers to report on severe storms. By 1887, 2,403 of them were on the job nationwide.

Finley was convinced that tornadoes often occurred where contrasting air masses met and that they favored the southeast quadrants of low-pressure systems. He correlated tornadoes to sudden shifts in humidity and pressure as well as wind. He also knew that tornadoes were more likely when a trough of low pressure stretched north–south (along what we would now call a cold front or squall line) rather than east–west. All of this thinking was way ahead of his time—30 years before such features figured prominently in the Norwegian storm theory.

On February 19, 1884, an estimated 800 people died in the Southeast in an outbreak of at least 60 tornadoes. A few weeks later, Finley took the momentous step of attempting daily tornado forecasts. Finley made forecasts based on the idea that when a large cyclone moved cool air over warmer surface air, creating instability, tornadoes were likely. After ending the experiment in August, he announced a 96 percent success rate. Other scientists took Finley to task for his statistical sorcery. Finley scored it a success if he didn't call for a tornado and none occurred. By that method, he would have had a higher success rate always forecasting no tornadoes.

Gustavus Hinrichs, the Iowa professor who documented derechos (the damaging straight winds of thunderstorm lines), thought Finley's claims were outrageous. An ardent booster of the salubrious climate of his home state, Hinrichs insisted tornadoes struck Iowa only twice a year. Hinrichs believed Finley counted derechos as tornadoes and labeled him a "professional tornado manufacturer." But Hinrichs' numbers were low because he didn't count tornadoes individually—only whole tornado outbreaks. Neither man was counting on the high side. In 1991 Iowa got 53 tornadoes, 17 of them F2 or stronger. Just to show that was no fluke, 51 tornadoes struck the state in superwet 1993. Regardless of accuracy, one of Finley's greatest contributions to meteorology was to show how common tornadoes are. He documented more than 100 a year after 1880. As more people settled the interior of the country, meteorologists heard about more tornadoes. With a denser population, and more and more Doppler radars, the National Weather Service has recorded more than 1,000 tornadoes each year since 1990. Every state, including Alaska, has had a tornado. In fact, some areas of southern California get as many tornadoes per square mile as Oklahoma, although, admittedly, these are weak twisters that often develop first as waterspouts over the ocean.

Finley's biggest headache was not Hinrichs but instead his own assistant, Professor Henry Hazen (no relation to the general). Hazen, who had studied under Elias Loomis, joined the Signal Service research office in

1881. To his credit, the professor pointed out that thunderstorms can speed ahead of their parent low-pressure system. He also showed that these storms quiet down at night yet begin anew the next day in the same sector of a parent low-pressure system.

Nonetheless, the enmity was obvious between these supposed collaborators. When Finley published his report on tornadoes, showing the importance of the southeast quadrant of a low-pressure system, Hazen publicly claimed to have made this insight earlier. The typos in Finley's original report contributed to this dispute, and eventually Hazen ceded the discovery to his boss. Hazen also doubted that tornadoes rotate, yet Finley insisted that nearly all tornadoes rotate counterclockwise. While Finley did his best to disprove electrical theories of tornado creation, Hazen believed birds lost their feathers in twisters because of electrical charge. Unlike Finley, he maintained that temperature boundaries play no part in tornadoes. Both Finley and Hazen were partly right on this point: the temperature contrasts Finley studied were on a scale too small to notice

The base of a supercell thunderstorm lowers a menacing wall cloud, often the parent formation of the most violent tornadoes. Rick Schmidt/*Weatherwise* photo contest.

on the synoptic maps Hazen and other forecasters focused on. Many years later, mesoscale studies would resolve this conflict.

Finley eventually left the Signal Service. He mustered considerable political connections to get a measly promotion to Lieutenant before his career came to a dead end. He became a leading figure in the insurance field, a founding member of the American Meteorological Society, and, in the 1930s, an expert on aviation meteorology. The intense "brain work" did him no harm: he died at age 89 in 1943, still actively teaching meteorology back home in Michigan.

Meanwhile, the Signal Service backed away from tornado forecasting and research. In 1887 a new chief shut down Finley and Hazen's office. Pressure from the civilian movement to wrest control of the weather service from the military figured in the decision. Even worse, however, General Adolphus Greely banned the word tornado from daily weather advisories. General Hazen first decreed this in 1883, when tornado forecasting seemed fruitless and misleading, but had reversed himself briefly during Finley's project. When the Agriculture Department took over the National Weather Service, the ban stuck. Thousands would die of tornadoes without warning over the following decades, but the meteorological brass was reluctant to risk discredit or to create a panic with half-baked forecasting techniques. The ban stood until World War II.

Tornado studies ground to a halt in the United States during this period. No one renewed Finley's tornado statistics until 1916. In fact, the Weather Bureau effectively stopped researching large hail and other effects of strong thunderstorms. European scientists recognized fully the irony of carrying on tornado research without valuable data from the country that has, as it turns out, perhaps three-quarters of the world's tornadoes and nearly all of the most violent ones. Alfred Wegener, who explained rainfall from ice crystals, led the field early in the century. In fact, his visit to Estonia in 1918 inspired Letzmann to study tornadoes. Wegener said that tornadoes came from rotation in thunderstorms. If wind shear created giant tubes of horizontal rotation aloft, the thunderstorm updraft could bend the tube. One end of the tube caused the tornado, while the other most probably caused an anticyclonic vortex.

For many years, Letzmann tried in vain to verify Wegener's hypothesis by recording observations of horizontal rotation in the clouds. But no reports of horizontal rotation withstood scrutiny. Letzmann also never found the pairs of counterrotating vertical vortices implied by Wegener's idea. Nonetheless, the great German meteorologist was on the right track

by relating shear to rotation in severe thunderstorms. Some of his predecessors, like William Ferrel, had insisted instead that a tornado rotates
solely because the Earth rotates. They said a twister concentrates a broad,
cyclonic wind field at the surface. This concentration of rotation obviously
exists, but on a much smaller scale than Ferrel envisioned. The rotation of
the earth only affects vast circulations like large cyclones, hurricanes, and
MCCs.

In the 1930s, armed with the new Norwegian cyclone theory, some
American scientists reinvestigated the relationship between low-pressure
systems and tornado outbreaks. They elaborated on Finley's findings, with
modern terminology. One study showed that nearly all tornadoes occur
near a surface cold front. Another showed that tornadoes touch down
ahead of the front. Finley and Letzmann believed that cold air aloft was a
key to tornado formation; many scientists began to agree. Cold air aloft
moving ahead of the surface front could cause considerable instability,
they reasoned, perhaps triggering prefrontal squall lines and thus tornadoes. This prevailing theory of cold air aloft got shot down in 1950 by
Chester Newton, a University of Chicago student of jet stream guru Erik
Palmén. Newton showed that squall lines ahead of the cold front were a
meeting between slow surface air and high-speed currents descending
from jet levels.

By this time, the amnesia over severe thunderstorms had ended at the
U.S. Weather Bureau. During World War II, munitions plants sprouted up
all across the country. These hazardous facilities made military planners
nervous. One lightning bolt in the wrong spot could kill hundreds of
civilians. Dense networks of volunteer reporters near the munitions plants
began tracking thunderstorms for the Weather Bureau. By the end of the
war the bureau had 200 networks, each covering a 70-mile-wide area.
Censorship prevented publication of weather reports, but tornadoes were
unusually frequent in 1942. The next year the bureau began issuing warnings to the public when observers spotted tornadoes.

During the war meteorologists also began to plan the famous Thunderstorm Project. This study didn't help scientists understand tornadoes—
the garden-variety storms studied in the project in Florida and Ohio work
differently from the severe storms that spawn the strongest tornadoes and
large hail. Weather Bureau scientists weren't actively investigating this
distinction. The U.S. Air Force, fortunately, was doing just that in the 1940s.
More specifically, Major Ernest Fawbush at Tinker Air Force Base was
looking for ways to forecast severe storms (which by definition produce

three-quarter-inch hail, flash floods, tornadoes, or other damaging winds). In 1948 an emergency forced him to put his ideas to work in a hurry. It was literally trial by fire.

On the evening of March 20, Fawbush's new assistant, Captain Robert Miller, settled into forecasting duties at the Oklahoma City area base. He studied the latest charts. Other than gusty low-level winds, the night looked like it would be calm. He issued a warning for 35-m.p.h. winds to hit at about 9 p.m. That was all. At that hour, weather observers began reporting thunderstorms to the west. Miller was surprised. The radar showed these were no ordinary storms. They were quite violent. Miller and a sergeant on duty prepared a new forecast. But it was too late. A squall line swept across Will Rogers Airport, seven miles away. The report: 92-m.p.h. winds and a tornado—all heading for Tinker. "We could only pray that this storm would change course ...," Miller later wrote. "There was no such miracle." The native of southern California watched unbelievingly as the tornado sped onto the airfield at close to 50 m.p.h. Miller ducked before the windows of the weather office burst and debris flew overhead through the room. The forecast was blown—literally. By the time the tornado lifted at the edge of the base, the Air Force had lost 32 very expensive airplanes, not to mention the damaged buildings. The toll exceeded $10 million in valuable 1948 currency.

The next day the Air Force declared the disaster an "Act of God," but the base commander ordered Fawbush to develop new techniques for forecasting tornadoes. Fawbush was ready. He and Miller spent several days reexamining weather maps for tornado outbreaks and produced a list of common conditions for severe weather. On the morning of the 25th, they noticed that the conditions of the first tornado seemed to gather again. They projected a severe weather outbreak over central Oklahoma between 5 and 6 p.m. Miller later admitted that the chance of two tornadoes in less than a week seemed about one in 20 million. But that's exactly what happened. Fawbush and Miller watched nervously as a squall line formed at 2 p.m. The line seemed to fizzle out: at Will Rogers Airport at 5 p.m., the storms had only moderate winds and pea-size hail. Miller went home disappointed. Then, over the base, two storms in the line seemed to rotate about one another and merge. In the heavy rain under "greenish-black" skies, a funnel suddenly descended to the ground. In less than five minutes, the base suffered another $6 million in damage, but this time precautions based on the forecast helped considerably. No one was hurt.

Fawbush and Miller forecast severe weather when several key features coincided in one area of the weather map. To guarantee extra instability, they looked for a narrow stream of low-level warm moist air, usually moving north from the Gulf of Mexico, and dry air entering from the west at higher layers. This usually ensured that updrafts would be warmer than their surroundings as they rose. Then the forecasters checked for a cold jet stream angling in over these winds. At this point all a region needed was a trigger to start the warm air rising into the dry air above. The trigger could have been a cold front or a trough of low pressure high in the troposphere. The severe storms were then likely near the intersection of the edge of the moist tongue of air below and the jet stream above. The need for strong triggering boundaries or swift jet streams ensures that tornado season begins in earnest by April and ends by late June in the Midwest and Plains (though somewhat earlier in the Southeast). Typically, half the nation's tornadoes strike during these months, though tornadoes have occurred on almost every day of the year.

Forecasters basically still use Fawbush and Miller's techniques to issue tornado watches. Their methods owe much to Finley—for instance, that tornadoes usually form southeast of a low pressure center in unstable warm, humid air, and that the severe parent storms form along boundaries like fronts or prefrontal squall lines. At the same time, Fawbush and Miller showed that tornadoes form in an area with great directional wind shear from the bottom of the atmosphere all the way to the top of the troposphere. The moist tongues enter from the south or southeast, the dry air from the west, and the jet often from the southwest. It's amazing that the atmosphere doesn't twist itself into a knot every time such conditions exist.

Wegener would have been pleased, to say the least. Forecasters now were looking for shear on a synoptic scale; soon researchers proved the significance of shear on a much smaller scale—that of the thunderstorm itself. The idea developed over the course of about a decade, beginning with the Tornado Project in 1953. Weather Bureau storm researcher Morris Tepper initiated the project to test his own theory about where twisters form. Tepper knew that fine-scaled analyses like Byers' and Fujita's showed that thunderstorm outflows increased pressure locally at the surface. Most forecast map makers ignored these slight detours in the pressure contours of a large, synoptic map. Tepper, however, believed the leading edge of this outflow was significant. It could wedge under the air ahead of it like a cold front. He called this minifront a "pressure jump" line, and he believed that the intersection of pressure jump lines near a squall line could trigger tornadoes. In the Tornado Project, Tepper and Fujita analyzed data from a special network of meteorological instruments in Kansas, Nebraska, and Oklahoma. Some tornadoes formed at the intersections of pressure jump lines, but others did not. The project was not a disappointment, however. Fujita and Tepper saw that the high-pressure pool formed by the thunderstorms had a low-pressure mate. This meso-low is an even better indicator of tornadic thunderstorms than pressure jump lines.

Mesolows, or mesocyclones, had been suspected for many decades. Then, in 1948, Edward Brooks of St. Louis University had the good meteorological fortune to gather data on four tornadoes (and three other funnels aloft) that passed within 30 miles of his home. Several occurred in low-pressure areas only 10 miles across (the small lows that crossed the Tornado Project network were 30–40 miles across; later scientists refined the mesocyclone definition to rotation that appeared to be about two to six miles across in the thunderstorm). Brooks showed that these mesocyclones were like the synoptic low-pressure systems that formed fronts. Mesocyclones are little wavelets on the larger cold front, and they wrap cold and warm air around each other locally as if in a miniature occluded storm. Brooks noted that in one case, the mesocyclone winds reached 50 m.p.h. People thought the tornado had widened to amazing proportions, but instead the mesolow had contracted and intensified.

Another thunderstorm fooled people in the same way as it approached Fargo, North Dakota, on June 20, 1957. That afternoon, underneath a towering thunderstorm, a menacing, two-mile-wide rotating cloud cruised eastward along U.S. highway 10. Frightened motorists managed to speed ahead of the huge cylinder to warn authorities in Fargo of the impending disaster. Broadcasts relayed the message, but instead of taking shelter, many people took advantage of the ample warning time. They pulled out their cameras and waited for the tornado. When the storm arrived, it was deadly, claiming 10 victims, injuring more than 100 people, and destroying or damaging some 1,300 homes. But because of all the pictures, the Fargo tornado of 1957 still ranks as one of the best-documented tornadoes in history. As a result, it became a turning point in tornado science.

Horace Byers heard about the photos and told Fujita. The master tornado detective gathered 150 photos and movie clips from 30 different

University of Oklahoma scientists brace their portable Doppler radar to measure the winds in a tornado near Hodges, Texas. Photo ©1989 Howard Bluestein.

photographers. For two years he poured over the images, calculating the azimuth of each shot and adjusting distances for the focal length of each lens. Fujita called the rotating protrusion from the storm a "wall cloud." In Fargo this rain-free cylinder extended halfway to the ground from the 3,000-foot-high cloud base. But near the highway outside of town, people had mistaken it for a two-mile-wide tornado because the huge lowering was very close to the ground and spinning fast. Fujita's analysis showed air rising at close to 50 m.p.h. near the top of the wall cloud on the north side. In other words, the wall cloud marked the intense core of the storm's rotating updraft, a concentration of the greater mesocyclone. The mesocyclone of the Fargo tornado in fact continued rotating for more than three hours, moving at nearly 20 m.p.h. from Buffalo, North Dakota, to Dale, Minnesota. Fujita showed that five different tornadoes had formed along the edge of the wall cloud; the Fargo killer tornado was third, dropping from the south side, away from the storm's rainshaft.

The wall cloud, sign of an intensifying mesocyclonic updraft, has become a familiar signpost for storm chasers. When they see one, they know they are on to a powerful storm with great tornado potential. In the 1950s, meteorologists also found evidence with radar that severe thunderstorms rotate with or without a tornado. The equipment—mostly World War II leftovers—was crude, making scientific evaluation difficult. Nonetheless, the thunderstorms' signals looked unmistakably different for severe weather. While tracking tornadic thunderstorms in 1953, meteorologists with the Illinois Water Survey in Urbana and with the Weather Bureau in Waco, Texas, and Worcester, Massachusetts, saw curled tails stretch from massive thunderstorm echoes on their scopes. They called the signals "hook echoes." (In 1944 a military radar operator in Alabama saw the "6-shaped" signature while tracking a tornado in Alabama, but he didn't publish the observation.) Somehow the precipitation was wrapping around an area without large drops. Storm-scale rotation appeared to be the cause—another verification of the mesocyclone dominating severe storms. About half of the hooked storms seemed to spawn tornadoes.

Not long after Fargo, an intensive radar study in England led to the next logical step: a new theory of rotating thunderstorms. An international team of scientists, including Frank Ludlam and Keith Browning, aimed five radars at a hailstorm over Wokingham in 1959. The storm cloud reflected the radar beams up to 45,000 feet, but at lower levels the echo was remarkably narrow. The radar echo in the center of the storm created a giant, relatively rain-free cloud cave with an arched ceiling 16,000 feet high and 25 miles wide. In the front of the storm the cave was open to about 12,000 feet. Updrafts sucked into the front of the storm under the overhang shot upward through the cave so fast that no large drops formed until the air reached the ceiling. In the usual (nearly) horizontal radar cross section, the cave was the empty area inside the hook.

While Browning was looking over the data from the radars at Wokingham, Ferdinand Bates of St. Louis University was photographing storms in the American Midwest, gathering powerful visual evidence of leaning updrafts in the strongest thunderstorms. Bates believed that leaning updrafts separated the strong thunderstorms from the weak, because rainy downdrafts would settle outside the energetic core of the storm. In 1962, the government's National Severe Storms Project sent Air Force planes into storms with leaning updrafts. The air inside rose at close to 150 m.p.h., more than twice the greatest updraft speed clocked during the Thunderstorm Project. Bates' ideas about rotation began to check out, too.

In the same year, a radar showed a tornadic storm over Kansas City with a hollow hoop echo that Bates had predicted could result from the meso-cyclone. High-altitude reconnaissance also revealed rotation in the updraft dome overshooting the anvil.

Browning, who participated in the National Severe Storms Project of the early 1960s, eventually synthesized radar findings and eyewitness accounts. He proposed that a severe storm is sometimes a "supercell" featuring a steady-state rotating updraft. This highly organized storm is the perfect parent cloud for a violent tornado. It begins its life as a fast-rising cell of convection, like the usual thunderstorm. But rather than housing a collection of equal convection cells at varying stages of development, one rotating updraft and its companion downdraft dominate the other cells. Indeed, a classic supercell has only one cell—one updraft and one downdraft. The sides of the towering cloud ripple with rotation, like a corkscrew or a barbershop pole. These storms can last more than six hours and generate updraft speeds exceeding 150 m.p.h. Since the low pressure at the heart of a twister demands a powerful, long-lasting geyser of air to act as a safety release valve from above, the discovery of the supercell thunderstorm energized tornado research.

The powerful updrafts also help explain how a supercell like the Dallas–Fort Worth storm of May 5, 1995, can create large hail. But not without complications—the main updraft is usually too strong for the initial hail embryos to form. The storm spits the ice out into the anvil. According to Browning, crystals grow better in the milder updrafts around the back (western) edge of the main cell. The crystals rise and rotate around the southern edge of the mesocyclone and form graupel in an icy margin called the "embryo curtain," the same curtain hanging over the mouth of the cave shown in radar echoes. The embryos spill out of the curtain and into the entrance to the cave. The powerful updrafts sweep the small hailstones toward the roof of the storm. Here the droplets around the hail are tiny, but they are also numerous. In fact the hail is in a perfect environment for growth. This model of supercell hail, vastly different than the Soviet model, in which the hail bobs atop the main updraft, is itself disputed. In another supercell hail model, the embryos form in a line of smaller, fast-growing cumulus that often flank a supercell to the south, along the outflow boundary. Supercells are hungry, long-lasting storms, which makes them bad parents. Like the mythical Saturn, they devour their children. The growing clouds beside the supercell, along the flanking line or gust front, get sucked up into the main updraft.

The supercell feeds itself with a constant diet of warm, moist air plowed up by its outflow boundary, the gust front. This is true of ordinary thunderstorms too. But a supercell gust front scoops up the inflow incessantly, vigorously. You can see this happening as puffy but underdeveloped scud clouds skim low over the landscape in the moist air (just like Roy Hall saw in the southerly air ahead of the tornado). The small clouds then sweep up over the gust front into the mouth of the waiting supercell. The lip of the storm curls up into a snarl: as the gust front or cold front clashes with the warm inflow, its upper surface curls back, rising enough to form what chasers identified as a "roll cloud" along the leading edge of the updraft base.

Supercells develop these special features because they form in different conditions from most thunderstorms. As Fawbush and Miller discovered, wind shear—the change in wind direction with height—favors severe weather, and this is the key to Browning's supercell. Shear is so important that forecasters today check for severe weather potential by comparing the strength of shear with the instability of the temperature strata. Relatively strong shear can lead to a supercell, but if shear is too strong relative to the instability, then storm cells get torn apart too quickly. (The exception to this is in a hurricane, where shear forces air to converge so violently that it creates uplift.)

Without wind shear, if one ailment doesn't kill the storm, another will. The inflow is timid; as a result, the gust front pushes ahead of the storm too quickly, denying the storm its fuel. Or the thunderstorm dies because the downdrafts fall right into the updrafts, snuffing the storm like water on a fire. In the shearing environment that supercells love, the inflow puts up a stiff resistance to the gust front so that warm, moist air jumps up right into the thunderstorm cloud rather than too far ahead of it. And in a shearing environment, the changing winds with height tilt the updraft, so that precipitation falls outside it. The rain-cooled downdraft stands by its updraft. As the precipitation falls, the mesocyclone wraps the drops around the outer edge of the updraft, creating the hook echo. In the meantime, Browning's model shows midlevel dry air wrapping into the supercell, helping to evaporate the precipitation and form an enhanced cool outflow at the surface.

Wind shear comes in another flavor besides veering, the clockwise turn with height that explains the wrapping downdraft and tilting updraft. The other shear is varying speed with height, and this helps explain the origins of the mesocyclone itself. Winds a few dozen feet up are faster than those at the surface—and so on up to the Indy 500 speeds of the jet stream.

This speed shear helps roll up the inflow air along the ground into tubes. So does the shift in direction—the low-level nudge from the southeast and the higher nudge from the southwest are enough to spin the air along its generally northward progress. This direction shear works just the way John Elway's hand does—fingertips pulling one way while palm pushes the other—to throw a perfect spiral touchdown pass.

Then an updraft comes along and bends the spiraling tubes upward, creating twin updrafts that rotate in opposite directions. All of this is just like Wegener's mechanism, only the rolls begin near the surface. University of Oklahoma meteorologists simulated these horizontal helices in Neil Ward's old tornado chamber. The rolls got dragged into the vertical by the updraft just as theorists expected. The chamber then produced a miniature tornado inside the cyclonically rotating updraft.

The roll-and-lift model of supercell rotation left theorists with a problem similar to the one Letzmann wrestled with on Wegener's behalf. What happens to the anticyclonic side of the twin vertical rotations? The world's fastest computers eventually provided an answer. In the mid-1970s scientists began to confirm supercell mechanics with landmark supercomputer simulations. The modeling marked a new phase in severe storm research. Unlike lab models, supercomputers can show a whole storm. Similarly, because basic physical equations dictate the motions, scientists can trace the importance of various factors in the storms, especially as better and better models include tricky processes like precipitation formation. The computers tend to show what observers first see in the field. But in the computer, the storms can be altered and run over and over to isolate key moments.

Joseph Klemp of the National Center for Atmospheric Research and Robert Wilhelmson of the University of Illinois National Supercomputing Center set up their equations with a modicum of representation of the microscopic workings of clouds. They also could only sample their cybercloud on a grid of hundreds of points about a mile apart. Nonetheless, in a landmark 1978 article, they showed that the computer simulated the expected emergence of rotating twin updrafts, starting with real temperatures and pressures sampled during an actual tornadic storm. The computer cells had the staying power of a real supercell and moved like one too. They also sucked in dry air at midlevels, as expected. In the cyberworld, the rainfall-induced downdrafts helped split the twin rotation in about 45 minutes of real time. The split was something meteorologists had noted before.

At this point directional shear becomes destiny. Klemp and Wilhelmson's models showed that the typical wind shear weakens the northern (anticyclonic) updraft of the rotating twins. Only by twisting around the directional wind shear were modelers able to make the anticyclonic cell survive. The cyclonic survivor heads southeast toward the warm inflow. This helps ensure longevity, because moving into the inflow makes the inflow that much stronger. For this reason, storm chasers look for the "right-mover" among a line of storms. That's where the action is.

Altering the conditions in the model, Klemp and Wilhelmson got to prod the beast in a way no scientist could in real weather. They found that a supercell is remarkably resilient about its rainfall. Indeed, in real life scientists have found that some supercells generate very little rainfall. Many high-based storms in the drier air of Colorado produce no rain, but their cooled downdrafts have plenty of space to accelerate into fierce dry microbursts. Storm chasers have found that some low-precipitation supercells can make tornadoes. Supercells with prolific rainfall form more frequently in the Southeast. Many of them are so wet that their wall clouds are ragged or even obscured completely by the shaggy cloud base and rainfall. But they make intense tornadoes in those dangerous shrouds of rainfall. A powerful tornado near Huntsville, Alabama, in 1989 killed 12 people in cars as it crossed a highway. The victims didn't realize the tornado lurked within the heavy rain.

Interestingly, supercomputer simulations also showed that supercells will form and thrive for hours even on an Earth that doesn't rotate. Shear is far more important. But the supercomputer supercells were picky about other conditions. If the rain formed but didn't evaporate significantly by entraining dry midlevel air, the downdraft was weak and the supercell never developed. Interestingly, though, modelers also found that in simulation, the twin updrafts of the storm would have refused to split if shear had been either twice or half as much as in the original conditions.

Finding the basic conditions that make supercells has not been nearly so difficult as figuring out what it is about supercells that makes them prolific tornado producers. Both simulations and observations have shown that these are not the same questions. When Neil Ward died at the untimely age of 57 in 1972, the discovery of the supercell seemed to be pointing researchers in the right direction. With Doppler radar, scientists at NOAA's National Severe Storms Laboratories in Norman, Oklahoma, had a powerful new tool to study airstreams in supercells. And the day of Ward's funeral, April 14, ironically opened a new age for NSSL's tornado researchers. They were embarking on their first official storm chase—

With mathematical wizardry and advanced supercomputing might behind them, simulated tornadoes spin up where quickly intensifying updrafts stretch a narrow zone of shearing winds. The ultrahigh resolution model resembles real outbreaks of rows of landspouts on the High Plains. Courtesy of Bruce Lee/University of Northern Colorado; Robert Wilhelmson/NCSA.

inspired by Ward's example to get out into the field and record visually what the radar was finding. It was perhaps a high point in optimism about eventually solving the tornado puzzle. Scientific chasing, Doppler radar, and computer simulations quickly added greater detail to the picture, but the pieces didn't fall into place as easily as scientists might have hoped.

At first, the storm chasers were limited to chasing in Oklahoma, and there weren't many tornadoes in 1972 anyway. Then came 1973: the first year totals topped 1,100. The NSSL and University of Oklahoma chasers scored a big success on May 24, during a four-day outbreak of 190 tornadoes. Hunting for severe weather in west-central Oklahoma, they began watching a small low-pressure center form at 11 a.m. along a passing cold front (extending from the main low in the Dakotas). Initially thunderstorms sprouted up along the front, but then, at 2 p.m., an isolated thunderstorm ahead of the front and just north of the little low rocketed upward. Two teams of chasers were only 25 miles away from the storm; one moved along the south side, while the other began tracking the north and west flanks of the storm. An hour later, the storm neared Union City in central Oklahoma. The chasers were right beside it as the wall cloud began to form below the storm base. One team, intent on watching the constantly changing wall structure, almost missed a pair of funnels that formed and dissipated overhead, about eight miles northwest of the mesocyclone.

The wall cloud spewed funnel clouds for a while with no result, but finally, at 3:38 p.m., one smooth-walled tornado cloud touched down. The damage swath was an eighth to a quarter mile wide, even though the visible cloud of the tornado continually narrowed to much less than that as it intensified on entering Union City. After passing through Union City, the tornado began to look more like a rope flopping from the clouds, and the damage path, still intense, narrowed to less than a 100 yards, eventually vacuuming together a row of debris as if it were a suction spot. By 3:53, the tornado lifted and disappeared. With ample warnings, only one person died, even though the tornado leveled a small mobile home park and other buildings along its path through the small town.

Seven miles away, the scientists watched the atmosphere curl into the tornado at the edge of the wall cloud. The flanking line and shelf cloud stretched from the south, arced around the storm base, and approached from the northeast. A team of chasers north of the tornado had to stop short when the rain shaft in the downdraft proved too dense and riddled with hail. The hail fell sorted by size—the smaller stones being farther north and away from the tornado—probably because of the tilt of the updraft above.

Throughout the chase, NSSL Doppler radar scanned the scene. From far away, the radar beams could not resolve the tornado or the surface circulation, but the evolution of the cloud was fascinating. Chasers confirmed that nearly exactly when the wall cloud formed, the radar detected rotation between 16,000 and 30,000 feet—a mesocyclone at first three miles across. The vortex aloft narrowed as it intensified (reaching more than 50 m.p.h. when it shrank to less than a mile wide during the tornado). This focused circulation spread downward in the storm while higher up the rotation was more diffuse and temperamental. As one scientist put it, the radar suggested a "vast, wildly boiling cauldron above the smaller ...

more organized vortex." Meanwhile, a full 20 minutes after the meso-cyclone formed, a hook echo appeared at lower altitudes.

With chasers on the ground near the tornado, radar operators learned that a half hour before "touchdown," an intense, focused rotation formed on the edge of the mesocyclone. At about the same time, the weak echo cave in the updraft slowly separated from the rotation and began filling in, a sign that left- and right-moving updrafts had split. This split seemed to liberate the rotation center; it began to swirl faster. The small rotational blip on the edge of the mesocyclone began to descend as it moved closer to the center of the larger vortex. This was the moment the chasers saw the tornado.

The scientists studied radar images of other storms and discovered that the small rotations embedded in the mesocyclone intensified and descended from about 10,000 feet before eight of ten tornadoes. They called these "tornado vortex signatures." First the mesocyclone, then the TVS, and finally the tornado—the rotation at the ground seemed to be the culmination of a chain of cascading vortices. Early tornado warnings seemed possible if meteorologists could find the TVS, which stretches as much as seven miles into the air, high enough for distant radar to see into storms when the tornado itself may be over the horizon. Among other things, the TVS discovery spurred the development of new Doppler radar now deployed nationally to enhance storm warnings. The Union City storm chase also had proven the value of having trained scientists on hand when tornadoes strike. Without them, the radar operators would not have known the significance of what the instruments had recorded.

The Union City breakthrough helped end the days when storm chasers inhabited a fringe of the tornado science community. Howard Bluestein says that when he finished his Ph.D. at MIT and joined the Oklahoma faculty, he didn't know much about observing tornadoes. He'd driven around the countryside of New England looking at storms as a student in the early 1970s, but the flat Plains are an entirely different panorama of severe weather. "I'd heard rumors that some people in Oklahoma had seen some rotating clouds, but I didn't know that anyone had ever storm-chased," he says. Once he took a ride with the NSSL chasers, however, he was hooked, fully aware that "not many people considered it scientific back then. They thought we were just going out to have thrills and spills." Bluestein was one of the NSSL–University of Oklahoma group to help change that image. In the late '80s, Bluestein put a technical face on chasing by introducing portable radar. Wesley Unruh of

Los Alamos National Laboratories fine-tuned equipment that was light, battery-powered, and easy to set up and take down under stress.

The fine, 3-mm wavelength beam proved critical in identifying velocities in the small confines of the tornado. On April 26, 1991, Bluestein and his crew took their van out on the road toward the Oklahoma–Kansas border with an eye on instability and awesome, 60,000-foot-tall thunderstorms. Just as conditions there began to fizzle, new convection intensified in northern Oklahoma. The scientists quickly headed north, reaching the scene 35 minutes later, just as the storms were hatching funnels. One by one, new towers popped up south of the old ones. The hunting was almost too easy—one storm that fled the scene eastward produced an F4 at Andover, Kansas, killing 17 people. Bluestein eventually chose the southernmost storm. Baseball-size hail rocked the vehicle before the researchers maneuvered east far enough into the clear. There they saw "a huge black cylinder" rip across Interstate 35. When the van first stopped, Bluestein and his students didn't have time to set up; the tornado turned and began eating up the landscape between them. After quickly evacuating to a safer position farther south, near Red Rock, Oklahoma, the scientists tried again. This time they nailed it. Two men had to lean on the valuable piece of technological wizardry to keep it from being blown to Oz in the powerful storm inflow. As the tornado swept a house off its foundation a mile away, the radar clocked 284-m.p.h. winds—the fastest ever documented on the face of the planet.

Storm chasers meanwhile showed their value in the theoretical arena, as well. In 1981, Bluestein was following the progress of a thunderstorm in Texas. Not much seemed to be happening, but then he made a serendipitous discovery. "I turned around to talk to someone in the back seat," he says, "and spotted a tornado behind us." How could the chasers have missed it? Meanwhile, NSSL scientists studied the same storm on radar. They were surprised by news of the tornado, too, because the storm had no mesocyclone. Bluestein knew that waterspouts usually form in young, fast growing cumulus towers that lack rotation, so he called the tornado a "landspout."

The very name landspout indicates the confusion over what makes a tornado. Waterspouts are supposedly different because they form over water, but if they cross onto land, they count as tornadoes in the national statistics. Before Bluestein's discovery, strength—both of the vortex and the parent clouds—seemed to be as significant as origins in distinguishing waterspouts from tornadoes. In Benjamin Franklin's day, mariners believed they could destroy waterspouts by blasting them with cannon balls. The sailors knew they risked losing a mast and rigging, but only an idiot would have tried a similar tactic in the face of a fearsome tornado. In 1950 an airline pilot in Africa flew into a waterspout at 500 feet over Lake Victoria. The wall of water was so dense, he couldn't see the tips of his wings. The noise was "terrific." The plane looped sharply, pressing him back in his seat. Finally, 15 minutes later, at 15,000 feet, the pilot turned the engine off and the plane spilled out. The whirlwind wore the dope covering off the fabric of the old plane, allowing water to fill the tail section, but otherwise the pilot and his craft survived without a hitch.

One of the chasers at Union City, Joseph Golden, had gotten his start with waterspouts by accident too. As a graduate student in 1967, he and some meteorologist friends had flown a small plane to Key West for a quick vacation jaunt. Luckily the scenery there didn't inspire them to take many pictures. On the way back they had plenty of film left when they spotted a series of waterspouts. Before they ran too low on fuel, the students circled the towering whirlwinds, filming and photographing them. Golden returned to the Keys on a full-fledged expedition; he flew close to 100 waterspouts, and on the ground observers spotted another 200. Using smoke flares, researchers traced the airflow in and around the spouts, developing a model of the vortex development that proved remarkably similar to the evolution of the Union City tornado. Before the waterspout column is visible, the sea surface darkens around a light area, then the spinning airflow etches a spiral of alternating light and dark bands several hundred yards across. Similarly, a tornado often kicks up a whirl of dust off the ground before the funnel touches the ground, and the inflow spirals into the vortex the way the water bands below a waterspout. The Union City observations helped show that the surface inflow around tornadoes is not a uniform, symmetric convergence; the air organizes into a curling "jet" that wraps into the vortex.

In the 1970s scientists actually flew right through the waterspouts with instruments to measure the updrafts and temperatures—an unthinkable act in a tornado. The cores had updrafts up to 22 m.p.h., temperature rises of only about one degree, and pressure deficits one to two orders of magnitude less than the most powerful tornadoes.

Now, Bluestein's landspout had opened a can of worms in the tornado theory pantry. Other chasers documented tornadoes without mesocyclones, especially near Denver and elsewhere in the high Plains. In fact, chasers in Colorado identified so many weak tornadoes previously missed by radar operators, that in the 1980s the state doubled its record tornado totals of the 1970s. Like waterspouts, the landspouts seemed to favor the convection developing in between more mature thunderstorms. Their behavior suggested that they favor boundaries where horizontal shearing might occur. Radar studies showed this to be true: in addition to clashing outflows between mature storms, landspouts favor fronts and flanking lines trailing thunderstorms. Another important boundary conducive to landspouts is the dryline of the southern Plains, where dry air from the Rockies collides each day with Gulf of Mexico air. And Florida, with its frequent thunderstorm outflows and seabreeze fronts, is hospitable to landspouts too. A computer model by Bruce Lee, now of Northern Colorado University, recently confirmed the importance of preexisting boundaries in the lower atmosphere. Lee's model sprouted several landspouts on boundaries that met a friendly vertical shear at the right moment, just as chasers sometimes confront a half-dozen landspouts in a fruitful horizontal shear zone. Gust fronts occasionally form similar lines of vortices, but without convection overhead, these dusty gustnadoes are not really tornadoes, nor are they particularly damaging.

In the 1980s, landspout discoveries weakened the idea that supercell mechanics explained tornadoes. The supercell-less tornadoes showed that not all tornadoes need mesocyclones. At the same time, chasers had documented many mesocyclones that failed to make tornadoes. Initially, researchers thought that almost half of the mesocyclones detected on radar were forming tornadoes. By the 1990s, that ratio was sinking toward a quarter or less. The supercell and its mesocyclone clearly were prolific makers of tornadoes, especially the strongest tornadoes. But something else was necessary to make the low-level rotation that generated the tornado itself. The tornado vortex signature wasn't a sufficient explanation— sometimes the TVS itself seemed to spread from low levels upward, rather than from the mesocyclone downward.

In 1985, computer modelers Richard Rotunno and Joseph Klemp of NCAR used supercell simulations to show that a mesocyclone won't make low-level rotation if the rain-cooled downdraft doesn't form. This is because the downdraft creates two essential boundaries of cold air against warm inflow air. The first boundary is northeast of the mesocyclone directly beside the downdraft, and the second is south of the mesocyclone, at the leading edge of the sinking air wrapping around the updraft. These miniature cold fronts and the warm air moving along them into the storm create the spiraling necessary to make low-level rotation, the models showed.

The irony of the situation was clear to theorists. Now the supercell was seen as a huge apparatus that doesn't really create the tornado. It just sets up the proper minifrontal system near the surface necessary to make powerful tornadoes. Landspouts, by contrast, simply take advantage of a nonsupercell environment that already has similar shearing boundaries. Another computer-modeling team went so far as to suggest that environmental factors establish the all-important low-level spin long before the supercell arises. Simulations by Rotunno, and later, in 1993, with very high resolution by Lou Wicker (of Texas A&M) and Wilhelmson, showed that surface friction pushed the low-level environment over the edge, concentrating the last bit of spin into a focused vortex. Such studies could only now be attempted because of the extraordinary computer speed and memory necessary to resolve tornadolike vortices under the storm.

With such evidence suggesting that tornadoes are a near-surface phenomenon, not simply the inevitable result of the high-level rotation of towering supercells, tornado theory seemed to be unraveling at the seams. Perhaps the best one could make of tornado theory was that no two tornadoes—not even supercell tornadoes—form exactly the same way. Clearly, tornado theorists needed a way to sort out the low-level rotation mess. One answer was VORTEX—the "Verification of the Origins of Tornadogenesis Experiment" run in 1994 and 1995. This was no ordinary field project, but the largest and best-coordinated tornado chase in history. Whereas scientists had waited for the thunderstorms to come to their instrument network in previous experiments, this time the scientists took the instruments with them. The goal: to surround thunderstorms with a phalanx of meteorological instruments throughout the tornado-forming process.

By now meteorologists had the tools for the chase. VORTEX assembled an armada of vans and cars bristling with cameras and meteorological instruments. Some carried portable balloon launchers for setting free radiosondes quickly enough to catch inflow winds into the belly of the storm. Everyone was linked by radio and computer to the field commander, project leader Erik Rasmussen of the University of Oklahoma. With global positioning satellite and special software, Rasmussen could track precisely where his data came from for future use in computer models. The first major success of the project came on May 29, 1994, when scientists in NOAA's P-3 radar plane, usually used for hurricane hunting, flew by a tornado near Newcastle, Oklahoma. They ventured as close as four miles as the twister reached F3 strength, ripping bark off trees and tearing crops out of the ground. For the better part of a mile, the tornado caused damage on the ground before the funnel touched down from the cloud above.

The Newcastle storm was a full-fledged supercell, and low-level rotation accompanied the higher mesocyclone. The tornado itself stretched-up rotation of about three miles across into a cloud a half mile wide. But the circulation that spawned the tornado was limited to the bottom 2,000 feet of the atmosphere. Furthermore, the tornado formed on the flanking line, far from the midlevel mesocyclone. The scientists were surprised to see such a strong, long-lived tornado separated from a mesocyclone. One of the airborne observers, Roger Wakimoto of UCLA, had seen plenty of landspout tornadoes. Here, however, was a tornado apparently stronger than the usual landspout and accompanied by a rather broad, low-level mesocyclone. This wasn't a landspout. But it wasn't a classic supercellular tornado, either. The rotation around the tornado seemed to stretch upward, not down. It was a tornado that simply defied the easy categories. Without the P-3 so close by, it would have been hard to document the strange separation of the midlevel rotation from the tornado.

The next year tornadoes were far more plentiful on the Plains. VORTEX reaped a bonanza of data at hitherto unimaginable scales. First, in addition to the P-3, scientists began flying ELDORA, a modified Lockheed Electra prop plane developed by NCAR and a team of French scientists. ELDORA's dual Doppler system meant that for the first time one airplane could single-handedly scan the volume of the atmosphere and reveal three-dimensional wind fields. ELDORA made many fascinating, detailed cross sections of storms, showing the weak-echo cave and its overhanging hail-embryo curtain. It revealed curling outflows in dry air filled with moist insects reflecting the radar beams. When VORTEX chasers caught their strongest tornado yet, a June 8 F5 near Kellerville, Texas, both ELDORA and the NOAA P-3 were on hand. Bluestein and Wakimoto watched as the incredible twister sucked up asphalt from the ground only 1,000 feet below them. The radars revealed a tall, thin tube of rotation with a narrow eye thousands of feet above the tornado. This tornado vortex signature showed a record-setting rotation of about 150 m.p.h. high in the storm.

The tornado itself, of course, reached even greater velocities at the ground. This was the domain of yet another new tool, the DOW radar chase truck built by Josh Wurman and his University of Oklahoma colleague Jerry Straka. Wurman says the DOW technology is nothing especially new. The radar hardware is military surplus, and the computers in the van are basically the over-the-counter variety. But with its narrow "pencil-beam" capabilities, the DOW complements Bluestein's portable radars by boasting enhanced ability to penetrate storms at close range. In VORTEX, DOW resolved the spiraling inflow into the tornado and the eye within. The observations suggested the downdrafts inside, just as tornado modelers had discovered earlier with their miniature simulations. The DOW even saw how the tornado centrifuged debris out of the funnel, sorting matter in the vortex by weight.

Now with two DOW trucks and updated radars, Wurman and Straka are part of the vanguard of tornado chasers challenging supercomputer modelers and others to keep the theories rolling forward. A year after VORTEX the two trucks simultaneously scanned a tornado—the first dual Doppler data set of the complete life cycle of a tornado, including the inception of low-level rotation. The flood of new data will undoubtedly help meteorologists sort out the confusion of defining tornadoes and recognizing the essential factors in their formation—with or without supercells. Meanwhile, Wurman isn't content to merely take the radar out chasing on the Plains, as he did near Spencer, South Dakota, in 1998. The fine-scale radar observations promise to help sort out mysteries in other types of severe weather. Earlier in 1998, DOW was parked along the California shore as wave after wave of intense Pacific storms, brewed in an El Niño year, exhibited amazing near-ground jets of high speed winds. And on the Atlantic coast in Hurricane Fran of 1996, Wurman's team was on hand to record bizarre waves of turbulence, a "washboard" pattern like the low-level rolls that feed supercell rotation. Only chasing hurricanes is an entirely different matter. In Fran, the DOW was in the midst of the storm it was scanning. The hurricane is not one but dozens of powerful thunderstorms, merged together into the ultimate storm.

CHAPTER NINE

The Ultimate Storm

An omnivorous traveler, Hurricane Georges seemed intent on devouring as much as it could from a plentiful buffet of tropical islands on its cruise through the Caribbean and Gulf of Mexico in 1998. Thirty million people in 17 countries and three American states felt the coiled fury—the driving winds, the pounding surf, and the relentless rains. Many millions more watched reports of the storm closely and spent anxious days wondering what the hurricane would do next.

On Sunday, September 20, Georges' winds whirled around its center at 150 m.p.h.—pretty much as fast as any hurricane that spins in the Atlantic. Antigua, Barbados, and St. Kitts had already felt the storm's fury. Just a couple of months before tourists normally flock to these oases from snowy lands elsewhere, Georges dealt the hotels, restaurants, and marinas a catastrophic blow. The storm's sweet tooth appeared insatiable, taking a huge, irreplaceable bite out of the islands' lucrative sugar crops. One out of every 10 residents of St. Kitts was left homeless. Compared to the annihilation of St. Thomas during 1995's Hurricane Marilyn, which destroyed 80 percent of the homes there, the blow to St. Kitts was merely ordinary.

But Georges was not content with a brief Caribbean vacation. It swept westward across Puerto Rico, just as the terrible San Felipe Hurricane had when it killed 300 people there in 1928. The distant memory of the San Felipe disaster sent shivers down countless spines—and not just in Puerto Rico. The San Felipe hurricane later passed over Lake Okeechobee, in Florida. Its fierce winds raked across the shallow lake bed, building up a wall of water that smashed through weak earthen dams and rolled across farming communities nearby. More than 1,800 people drowned.

Puerto Ricans were far better prepared for the hurricane this time, but the cost was high. Georges knocked the power out for nearly everyone on Puerto Rico, and almost all of the island's four million residents were temporarily without running water. The entire island was drenched with rain (up to 20 inches in places) and lay under the threat of sudden flash floods. Three people died in a rain-triggered landslide. The damage toll approached $2 billion, including the ruin of at least 17,000 homes. But rebuilding began quickly in this U.S. Commonwealth, and surfers soon returned to the beaches.

Ironically, surfing was one of the few ways a person could have navigated the flooded island of Hispaniola after Georges was through there. The storm headed a bit farther north at first, seemingly ready to recurve poleward and then eastward, as practically all such storms do worldwide. But the semipermanent Bermuda-based high-pressure cell to the north—the airy east–west Great Wall of the Atlantic—stretched out long, straight, and impenetrable. Georges rebounded to the southwest with barely a moment's hesitation and swept across the Dominican Republic, the eastern nation of Hispaniola. The strongest winds, in the right front quadrant, raked the island while the eye moved along the south shore before plunging inland at San Pedro de Macoris.

The right side of the storm, as usual, also proved to be tremendously wet, as if the copious rains in Puerto Rico had merely whetted Georges' appetite for a good soaking. As much as 20 inches of rain fell in the lush mountains of Hispaniola on the night of Tuesday the 22nd. Slow to respond to the threat of disaster, the Dominican government opened shelters during the thick of the storm, too late for many citizens as the floods and slides began. In San Cristobal, Dominican Republic, the Nizao River overflowed its banks, drowning five people as the waters washed out part of a school being used as a local shelter. Dozens more people were missing. Others fled the rising waters that demolished their homes, grabbing empty water jugs to use as life preservers. They had time to save their children and little else, swimming in swift currents as their belongings floated by. More than 300 people drowned in the muddy overflow on Hispaniola, some of them buried forever beneath the silt and wreckage.

In the Dominican Republic alone, the cost of the damage easily exceeded the government's entire annual budget. Some people had no choice but to pry loose rusty nails to piece together what housing they could from the wreckage. Rescue workers dropping food from airplanes over the inaccessible mountain villages of the Dominican Republic saw the wide, flat swaths of mud that buried once-fertile valleys in the mountains north of the capital of Santo Domingo. Similar mudslides in Haiti may have been due in part to widespread deforestation, the local method for gathering charcoal.

The devastation left farmers without either crops or livestock—animals that in many cases constituted a life's savings. Only 10 percent of the crops survived in Haiti and the Dominican Republic, leaving the 15 million islanders with a rapidly escalating food crisis. Haitians suffered greatly from Georges, yet many there believed the storm could have been much worse. The attitude seemed naive to outsiders, but how else could locals compare the experience to Hurricane Flora of 1963, which killed nearly 7,000 people in Haiti and Cuba? As recently as 1995, mudslides from a much weaker Tropical Storm Gordon killed 1,000 people in Haiti. Indeed, the Haitians were correct. The similar, mountainous landscape of Honduras and Guatemala gave way under the incredible 50 to 75 inches of rain that Hurricane Mitch dropped in its landfall a few months later, killing at least 11,000 people in massive mudslides and widespread floods. Nearly every road and bridge in the region was washed out in places, and at least two-thirds of the crops in Honduras were ruined. Officials were not exaggerating when they reported that development of the flooded country had been set back nearly half a century during the storm. Mitch made 1998 the worst year for hurricanes in the Caribbean and Atlantic in more than two centuries.

Flora, Georges, the San Felipe, Mitch. Every year on June 1, the people of the paradisiacal tropical shores of the Caribbean, Atlantic, and Gulf of Mexico settle in for six months of anticipation, wondering when they'll have to relive the tragedies of previous hurricanes. The storms bring vital rains, yet cause treacherous floods. They blaze wavering paths that leave people gasping at improbable luck. From the moment it began forming hundreds of miles east of the Gold Coast of Africa on September 14, Hurricane Georges created a whole season's worth of miracle and woe and watery death in the Caribbean. Then it continued its exasperating journey in the Gulf of Mexico, giving Americans a glimpse into the disasters that threaten our own shores. Such typically long-lived, wide ranging storms test our resilience, returning to haunt the same places again and again.

For all our terrible experience with them, hurricanes continue to test the limits of our meteorology. In part this is because hurricanes test the limits of our credulity. No other storm marshals such vast resources so persistently. A conspiracy of powerful storm towers infiltrating a half million square miles, hurricanes get their energy from the sea. They are the only storms to fully mine the most plentiful resource on the planet—the solar energy of hot, tropical oceans. This cooperation between water and air makes these tropical cyclones the ultimate storm. Georges released enough heat energy in a few short weeks to power the electrical needs of the United States for a decade. Nor does any other storm spend its energy so frivolously—generating 10 times the heat it needs to gather a billion tons of air a minute with a sprawling coil of wind, then pump this extravagant horde of heat and moisture 12 miles straight up to the top of the troposphere. Such work could be performed only by a colossal mechanism, like an engine fueled by several thousand atomic bombs exploding every second.

Science has yet to give us a truly confident prediction of which clouds will form a hurricane. Nor can meteorologists be precise about where a hurricane will strike and how hard. Hurricanes, like all storms, are extreme events. They are the exceptional, not the likely, result of a bunch of tropical rain clouds borne of the same sweet tropical air and lazy warm water that generates oranges, sun tans, and shuffleboards. One thing about extreme events: they follow their own rules. Hurricanes quickly develop personalities that seem to defy their reputation for being cleanly organized, tightly wound spirals. They are fickle, erratic, headstrong. Above all, one would never dream of messing with a hurricane.

Like tornado chasers, hurricane specialists since World War II have met a rogue's gallery of these quirky storms and in the process have become acquainted with more bizarre tropical tantrums and seizures than they ever thought possible. The expanding repertoire of inexplicable behavior magnifies the immensity of the challenge of understanding hurricanes. The hurricane now poses problems for computer modelers from the very microphysics of the clouds to the dynamics of whirling eddies to the volatile jostling of a million cubic miles of circulation all the way to the interplay of planetary cycles.

At one time hurricane science seemed as simple as the observations were inadequate. Smart mariners quickly learned the rules of hurricane engagement. Columbus was fortunate to avoid stumbling onto a hurricane on his first voyage to the Americas. Having embarked in August, the height of the season, he was at least a little bit lucky in 1492. But in 1502 Columbus showed what he'd learned from the natives and his own travels. A bright, clear sky and strange westerly winds convinced him that a hurricane approached the Caribbean. He asked for shelter on the island of Hispaniola, but Ovando, the new governor there, didn't like the admiral and refused. So Columbus moved his ships to safety elsewhere and rode out the ensuing tides and winds. Meanwhile, scoffing at the warning, Ovando sent a treasure fleet off to Spain. Twenty ships went down in the hurricane, drowning 500 Spaniards and their slaves. Only one ship in the fleet survived; it carried Columbus' own fortune in gold.

William Redfield's survey of damage after the New England hurricane of 1822 led to the first thorough documentation of the circular pattern of winds in these tropical cyclones (which had been described as whirlwinds by various geographers for at least two centuries). Not long after, in 1855, Henry Piddington of India introduced the term "cyclone"—coil of the snake. This most apt of descriptions remains the word for these storms in the Indian Ocean and Bay of Bengal. It is also the generic term used by meteorologists, though perhaps it is no more apt than "hurricane," derived from the name of a Caribe storm god (and used in the Western hemisphere), or "typhoon," from the Chinese word for big wind that applies elsewhere in the Pacific. The coil Piddington conjured occasionally confused scientists into thinking these storms were overgrown tornadoes. In 1892, Ralph Abercromby noted that some of his peers believed that tornadoes could actually develop into hurricanes. Fifteen years later a meteorological textbook claimed that hurricanes and tornadoes differed only in size. The mistake wore away as tornadoes themselves were more firmly linked to specific thunderstorm updrafts. Nonetheless, it is ironic that storm chasers now gather super-fine Doppler radar images of a tornado that reveal a surrounding coil of winds that looks startlingly like a hurricane on a vastly reduced scale, including its trademark eye.

In the hurricane, spiraling, converging air from hundreds of miles away heads toward the eye. The low pressure in the eye is remarkable depending only on your perspective. It's trivial to anyone who has experienced the same pressure decrease simply by taking the elevator to the top of Chicago's Sears Tower or riding a ski lift in the Sierras. To a weather watcher at sea level who checks the barometer every day, however, 1/10th of an inch change in the roughly 30-inch-high mercury column is a significant event. The two- or three-inch plunge into an intense hurricane eye is nearly unbelievable, and the 25.69 inches in Typhoon Tip in 1979 is, well, a world record for sea level. A man who had ordered a barometer picked it up from his mail one day in September 1938 and took it to his beach home on Long Island. To his dismay, it read far below 29 inches and was dropping steadily. Thinking he'd gotten a dud, the man went to the post office to return the instrument. By the time he got back, his home was gone, dashed to pieces by a hurricane that caught the Northeast by complete surprise, killing close to 600 people.

The eye is a false haven for landlubbers. The Miami Beach of 1926 was a new city full of people recently transplanted from the North; almost none of the more than 100,000 residents had seen a hurricane. When winds suddenly stopped roaring during a hurricane in August that year, bewildered beachcombers wandered out to the water for a swim or to survey the damage. Forty minutes later, a new wall of wind and water swept over them, drowning more than 100 people.

Birds, on the other hand, often seek refuge in the eye. A flock of sea gulls gathered up by Hurricane Diana in 1960 circled for tortuous hours before alighting near Cape Fear, North Carolina. Meteorologists were able to track the flock as it moved hundreds of miles with the storm. Many of the birds dropped exhausted onto the water; their bodies drifted ashore over the following days.

On the surface of the sea itself, the eye is a deathtrap. One mariner who endured a China Sea typhoon in 1869 wrote that with "one wild, unearthly, soul-chilling shriek the wind suddenly dropped to a calm." Here the ship teetered atop mountainous, nearly conical waves that "boil and tumble as though they were being stirred in some mighty cauldron." The Confederate ship *Alabama* got caught in another while raiding Union shipping in 1863. In the eyewall, Commander Raphael Semmes wrote, "The tops of the waves were literally cut off by force of the wind, and dashed hundreds of yards, in blinding spray." The ship threatened to capsize, but then suddenly entered the eye itself. "The clouds were writhing and twisting, like so many huge serpents engaged in combat, and hung so low, in the thin air of the vortex, as almost to touch our mastheads." The peaked waves were now "cones" that "jostled each other, like drunken men in a crowd, and threatening, every moment, to topple, one upon the other." Scientists today who travel routinely into hurricanes see these waves and thank their knotted stomachs that they are flying, not sailing. Their view of the awe-inspiring stadium of sun-drenched cloud 50,000 feet on a side is a perfect case of the new perspective on storms. Edward R. Morrow, the CBS newsman, compared the sight to a bright blue Alpine lake surrounded by a perfect ring of snowy mountains sloping steeply right down to the water—"a great bowl of sunshine."

The seemingly unblinking gaze of the hurricane eye once cast such a strong spell on meteorologists that they became too enamored of the classic view of a hurricane as a perfect spiral—in the Northern hemisphere, a counterclockwise wind (about 10,000 feet thick) circling into the eye and a clockwise outflow of air above it. The form is so perfect that only a theorist could dream it up. That's why men like Espy and Ferrel, who identified this three-dimensional model, yielded to theoretical temptation and said that all storms had warm cores that pumped air up off the surface high into the atmosphere. Too many early storm theorists assumed that the ideal coil of a hurricane could happen anywhere, that it was the most likely of storm shapes.

In the 1920s and 1930s, the meteorological world briefly swung too far away from the ideal with a push from the successful Bergen school. The young scientists there drew such a convincing model of storms that other meteorologists began to look for fronts that might trigger tropical cyclones. At first glance, however, there's nothing like a front in the tropics. On rare occasions, the tail end of an extratropical front helps trigger a tropical cyclone. Sometimes, the convergence of winds between the hemispheres— the stormy intertropical convergence zone shifting near the Equator— spins off a hurricane. But this is not the norm everywhere.

Opponents of the invasion of frontal theory into tropical meteorology argued that hurricanes more likely formed the same way that isolated thunderstorms cropped up in the warm stagnant air southeast of an extratropical low. An errant warm spot in the uniformly warm atmosphere of the tropics fires up a towering updraft, they said. For instance, a sun-baked island often creates convective clouds when the ocean around it is generally clear. The edge of outflowing winds from Asia during the dry season also seemed to cause tropical disturbances. But at the time scientists couldn't figure out how this eastward-moving boundary—the monsoon trough—could create a westward-moving typhoon. (Today the shear and spin along the monsoon trough is recognized as the origin of many typhoons.)

Scientists grappled with the tropical cyclone problem without much evidence at their disposal. Before World War II, only a few weather stations represented the vast tropical oceans; at best, meteorologists worked with historical averages from far-flung stations; real-time analysis was unthinkable. But the global war changed that in a hurry, taking unprecedented numbers of meteorologists to the tropics. They soon confirmed suspicions about the insatiable needs hurricanes have for heat. The margin of error for hurricane formation is small, even in the warm, hospitable tropics. That is why hurricanes never form poleward of 30 degrees latitude. (They also rarely form Equator-ward of about 5 degrees because the Earth's rotation doesn't curve winds strongly enough there.) Only in the tropics are sea surface temperatures warm enough to provide easily evaporated moisture to grease the engine of the storm. Aside from this critical evaporation, tropical waters warm the inflow air by reradiating heat. Otherwise the air might cool too much by decompression near the intensely low pressure of the eye.

In 1947, at Carl-Gustav Rossby's invitation, the Finnish meteorologist Erik Palmén was visiting the University of Chicago, studying appropriately northern topics like the jet stream. He heard that a hurricane was heading for New England. Intrigued by the brute force of such storms, Palmén decided to make the most of the opportunity and went to Boston. The storm had other ideas, however; it changed course and came ashore in Florida. Disappointed, Palmén began to study tropical cyclones. He quickly saw that these storms are heat engines: they transfer the heat of the ocean surface to the top of the troposphere. The greater the temperature difference between the sea and the atmosphere, the stronger the hurricane can be. Palmén calculated that the water below must be at least 80 degrees Fahrenheit over a very wide area for the storm to crank its motor. This makes the tropical seas barely adequate. Sea surface temperatures there don't exceed 83 degrees except seasonally in a few select places. The water doesn't get any hotter because the steady evaporation at such high temperatures adds a strong compensatory cooling effect.

None of these ideas had a great effect on forecasting hurricanes. The tendency of hurricanes to strike when the water was warm was well known by accumulated experience. In fact, the leading hurricane expert of Ferrel's day was no theorist but instead a Jesuit priest with a keen eye. In 1870, Father Benito Viñes became head of the observatory at the Royal College of Belen in Havana, Cuba. Soon he established the hemisphere's first hurricane warning service, based solely on his careful observation of the clouds, tides, and the observatory's instruments. His exhausting regimen of observation included hourly temperature, pressure, and wind readings from 4 a.m. to 10 p.m. every day. Father Viñes issued a two-day advance warning of a hurricane that struck southern Cuba on September 11, 1875, which ensured his fame throughout the Caribbean. To further his work, the steamship companies guaranteed the Jesuit scientist free passage throughout the region.

Ironically, one of the first signs of a hurricane, according to Viñes, was the pure blue sky surrounding it (where clouds disappear in the sinking outflow of the storm). Even in this clear weather, Viñes could tell where the distant storm was. He simply turned and faced the direction across which gentle breezes blew left to right. At the appearance of the actual hurricane clouds on the horizon, Father Viñes added, the rain clouds "begin to

The deadly spiral of a hurricane was infamous centuries before meteorologists finally saw one whole, from space. NASA photo courtesy of Gregory J. Byrne.

overrun the skies with inexhaustible succession and high speed. Showers of short duration begin, and the wind velocity increases from that moment." The pressure, which was already falling slowly, now plummets, and the rain becomes continuous, "although highly irregular," as gusts begin topping 100 m.p.h. Before dense cloud cover closed in on him, Viñes watched the upper clouds, wispy cirrus, to ascertain the movement of the cyclone as a whole.

Father Viñes consulted storm-savvy residents of Cuba for clues to hurricane behavior. He reported that Cubans already knew well that hurricanes of the region move westward, and that as the season progresses the storms become a more serious threat. At mass, Cuban priests began praying "Ad repellendat tempestates" a month later than their brethren in Puerto Rico. In fact, in August and to an extent in September, hurricanes tend to form so far east that they often must hurdle Hispaniola before striking Cuba. Georges was one of these storms. It weakened during its climb over the highlands of Hispaniola, but it posed a significant threat to Cuba next, just like deadly Hurricane Flora of 1963. The Castro government took heed and evacuated some 200,000 people from the east coast. Only five died on the island as the storm once again scaled the mountains blocking its path to the Gulf of Mexico.

Long after his death in 1893, Viñes was the leading name in hurricane study. Inspired by his success, the United States began a hurricane warning service in 1898 in response to the impending war with Spain. The head of the Weather Bureau, Willis Moore, briefed President McKinley on the potential destruction an unanticipated storm could inflict on the fleet. McKinley exclaimed that hurricanes posed a more serious threat to the Navy than the Spanish Navy did. He agreed to establish a storm warning service on the islands of the Caribbean.

The system was not enough to prevent the worst natural disaster in American history. On August 27, 1900, a group of thunderstorms began to coalesce after passing the Cape Verde Islands as they rode the steady tropical trade winds westward. A week later the stormy area was a hurricane. It swept over Cuba and then the Florida Keys on Friday, September 5. The Weather Bureau began to get nervous about the storm and spread word to the Texas coast that somewhere, out over the Gulf, the hurricane might still rage. In fact, the *Pensacola*, a steamer bound for Florida from Galveston, was in the thick of the storm already. The crewmen were fighting for their lives in the huge waves. But the Weather Bureau had no knowledge of this, and most ships' crews that would have reported the progress of the storm wisely stayed out of the Gulf anyway.

Isaac Cline, the Weather Bureau chief in Galveston, began to worry about the storm too. Early on Sunday morning, he and other Galvestonians saw the waves pound the sand. The water slowly ate its way toward the beachfront structures. Cline knew the high water this early in the morning was not a tide, but a wind-pushed surge of water from the hurricane. Above, the clouds moved swiftly across the surge of water, not with it. The city was apparently already ensnared in the twisting circulation of the hurricane. Perched precariously on a sand bar off the Texas mainland, Galveston clearly had little margin of safety. Cline rode up and down the beach, warning people of the impending storm tide and urging them to flee to high ground. He knew that in a similar situation, the entire Texas town of Indianola had twice disappeared under a surge of water during a hurricane. In 1875, 176 people died and most of the houses were destroyed. After an 1882 hurricane obliterated Indianola again, the settlers gave up and never rebuilt. Cline realized that the fate of Galveston—the nation's second-busiest port at the time—hung in a balance poised recklessly in a hurricane.

By 10 a.m., the Weather Bureau in Washington was sure that the strong, right quadrant of the storm would hit Galveston. Two hours later the northeast winds exceeded 30 m.p.h. The east end of the city was under three to five feet of salty water. The train and wagon bridges connecting Galveston to the mainland across the bay were impassable. Only a few people had escaped in time. At 2:30, Cline and his brother Joseph sent a last message to Washington—"half the city now under water"—and waded back home to seal up Isaac's house near the beach. By 5 p.m. the city was completely submerged. People who didn't drown spilled out of their toppled houses into the hail of nails, glass, and wood borne by the 100-m.p.h. gusts. Within an hour the water had peaked at more than 20 feet above normal tide. Many neighborhoods lay under 15 feet of water. Cline's house, one of the best in the city, survived the initial rise of water. Inside, Joseph and Isaac frantically cut holes in the floor to keep the structure from popping off its foundation; then they raced upstairs with their four girls and Isaac's wife as the first floor flooded. But a cascade of debris rammed the house when the winds peaked at 120 m.p.h. and quickly shifted to the east. The house turned over into the waves, trapping Isaac's wife and younger daughter underneath. The rest of the Clines scrambled on top of the house and rode for three hours in the maelstrom, unaware of the fate of their loved ones beneath them.

That Sunday, without warning, the storm cost Galveston 6,000 of their loved ones. Perhaps 1,000 more people died inland. Galveston was ruined as a leading commercial city, but the city rebuilt itself on a new, thicker layer of sand, then constructed a seawall to arrest future storm tides. Fifteen years later, another hurricane made a direct hit and killed only eight people.

The seawall is perhaps the only real defense against the deadliest force the hurricane musters—the towering surge of water at and around its center. In the powerful winds of a hurricane, the surface of the ocean becomes a foamy green brew in which air and water are one. Nearly lost in the spray is a surface of water undulating with greater and greater force. The *Queen Elizabeth II* rode out 98-foot waves during Hurricane Luis in 1995. Other hurricanes routinely build up waves topping 50 feet. At the same time, the winds plow the water toward the storm's eye, gathering a massive dome of water.

The tumultuous waters in the eye mask a subtle rise in sea level over the deep ocean. The wind piles up the water and then holds it hostage in the storm's center. Even the low pressure of the eye contributes to the storm surge, raising the sea surface a couple of feet as if sucking it up with a giant straw. The mighty forces at work between air and water remain nearly imperceptible over the deep ocean. The only obvious evidence of the impending disaster is the train of waves emanating out from the eye, ahead of the storm. The booming, steady cadence of crashing surf is especially unsettling in the brilliant sunshine and calm preceding the storm.

It takes a cooperative coastline—a gently rising slope of the continental shelf—to unleash this water into a monstrous rise in sea level far above the normal tide. The water is continuously bucked up and down by wind-driven waves that raise the relentless inundation higher than a house. The increased friction along the shallow sea bottom forces the storm waves to grow higher and higher and bunch closer together. A steep shoreline, like the one ringing Jamaica, gives the storm surge little chance to pile up. By contrast, Galveston lies atop the gradual rise of the Gulf of Mexico shoreline. The same hurricane creating a 10-foot storm surge along parts of the U.S. Atlantic coast would raise the water more than twice as much near Biloxi or Galveston. Each foot is critical—the water crashes into the wall of a house with 1,000 pounds of pressure per square foot. And if water doesn't demolish the house by shear horizontal force, the overwhelming buoyancy of the structure will likely lift it off the ground and onto the waves, atop which the weakened structure almost inevitably collapses.

Geography has been singularly unkind to the people living near the wide river deltas ringing the Bay of Bengal. For many miles off shore the sea depth is only a couple dozen feet; the rise of the land itself is even gentler. One of the highest storm surges in history may have been in the Great Bengal Cyclone of 1737. A British settler reported that the sea rose 40 feet along shore and spread perhaps 100 miles up the mouth of the Ganges. The casualty total may have been 300,000, as reported, though this is unlikely as the population of the region was relatively low at the time. More conservative estimates show a death toll of 3,000. The salt water ruined farm land many miles inland. The tide did not ebb for almost a whole day. Every ship in Calcutta harbor disappeared, foundered, or broke apart; in all, residents lost perhaps 20,000 craft. Tigers and rhinoceroses, goats and chickens drowned along with people. A French crew in the area began to empty the soggy bales of cargo from their hold after the storm inundated their ship. One man went below deck to pass the goods up to the rest of the crew. After a short while he inexplicably stopped working and wouldn't answer his mates' calls. Two others went down to see what the trouble was. After they too, said nothing, the frightened men brought lamps down with them and found a very satisfied crocodile thrashing about. It had crept in through the smashed side of the ship.

At the time, the English could not believe a storm could ruin nearly every building in the region. In London, *Gentleman's Magazine* reported that an earthquake had shaken the ground during the cyclone, a misperception repeated in many other storm reports over the ensuing century. Sadly, today we have come to expect our greatest weather disasters to occur during the storm surges of the Bay of Bengal. The surge of the 1879 "bakerganj" cyclone rose 40 feet over the Bangladesh coast and killed 100,000 to 400,000. Even with a more "moderate" surge exceeding 20 feet, the November 12, 1970, cyclone killed more than 200,000—some claim as many as 500,000. And on April 29, 1991, Bangladeshis faced yet another "day of judgment," as one survivor called it. A cyclone storm surge 20 feet high raced dozens of miles inland, killing 139,000 people. The water rose relentlessly over the homes of temporary workers living on the shifting mud of the Ganges delta. Hundreds of people died in one of the few concrete storm shelters when the water burst through the door. On another low-lying island, only 650 people made it to shelter. They survived, while 2,000 of their neighbors were swept away around them. The waves destroyed more than 800,000 homes and damaged a million more. In many areas, the land simply disappeared; helicopters bearing food and other supplies couldn't even land. For those who survived, the disaster made farming nearly impossible with salty topsoil and the loss of a half million precious domestic animals; fishermen had no boats left with which to ply their trade.

Similarly unprotected lands lay ahead of Hurricane Georges when it left Cuba, already recurving northward, safely west of the Bermuda high. The storm that killed more than 400 people by rain and mud in the Caribbean now threatened a grim fate for the Florida Keys. As the hurricane approached this 100-mile-long chain of islands on Wednesday the 23rd, a steady stream of cars choked U.S. Route 1, the only two-lane evacuation route to the mainland. More than 100,000 people headed toward Miami under mandatory evacuation orders. Some of them undoubtedly thought back to one Labor Day more than six decades before when another hurricane made the Keys synonymous with tropical disaster.

In 1935, thousands of World War I veterans were marooned on the Keys in the strongest hurricane ever to hit the United States. The "bonus marchers" had descended on Washington by the thousands and pitched tents on the Mall, desperate for assistance during the depth of the Great Depression. The Roosevelt Administration finally answered the crisis by shipping the demonstrators down to Civilian Conservation Corps camps on the Keys. The men were nearly forgotten when a tropical cyclone loomed in the Caribbean early on Labor Day, September 2. Convinced of impending disaster, the Weather Bureau wired a warning to the Keys, advising the veterans to prepare for a relief train that would arrive in a matter of hours.

The bonus marchers, most on Matecumbe Key, waited along the tracks for the train to take them to safety. They expected it to arrive at 4 o'clock, but the tracks were empty. Instead the shattered remains of trees and shacks flew in the air and the water rose around the men. Finally the train appeared—too late. A wall of water more than 15 feet high dashed the remaining shacks to pieces. The train toppled off the tracks. The veterans had no choice but to ride out the storm and its 200-m.p.h. winds.

The eye moved slowly over the Keys and probably tightened considerably as the storm whipped itself to peak strength. In the middle of that tragic night, the stars shone nearly 40 minutes in the eye at Matecumbe Key and almost an hour at Long Key. At the time 20 feet of water washed over the men. Few structures could withstand the onslaught. Several dozen men clung to the top of a 25-foot-high tank car loaded with water. Shells flew like shrapnel in the air, cutting the men's faces as stinging saltwater waves crashed over them. The next morning they found some of their buddies hanging from the trees, the bare branches piercing their bodies. More than 400 people died on the Keys that night. One survivor swore, "I'd a whole lot rather be on the battlefield amid machine gun fire than go through such a storm again." An enraged Ernest Hemingway lashed out in an editorial, suggesting that it was manslaughter to send people out to the Keys during hurricane season in the first place.

The few who remained on the Keys as Georges approached were there by choice, hoping they might save their property. But most residents didn't take any chances. They didn't need to. For nearly two weeks NOAA's GOES-8 satellite trained its gaze on Georges. The satellite, which orbits the Earth at exactly the speed necessary to hover over the Atlantic constantly, gives space age forecasters a huge advantage over their predecessors. Around the world today, meteorologists track about 100 tropical

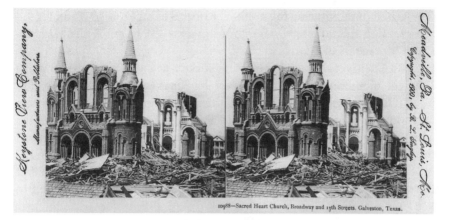

10988—Sacred Heart Church, Broadway and 13th Streets. Galveston, Texas.

Galveston's Sacred Heart church, reduced to rubble in the worst natural disaster in American history—the September 1900 hurricane that killed more than 6,000 people. Library of Congress photo courtesy of Patrick Hughes.

cyclones a year. Before the era of satellites, they were only able to find about half that many. The life-saving satellites are thus the single most critical hurricane-watching device we have, according to former National Hurricane Center Director Robert Sheets.

The dream of weather surveillance from space neared fruition after World War II, when Rand Corporation scientists began a classified project to study meteorology from orbit. Using captured German V-2 rockets, scientists began taking photographs of clouds from space, retrieving the film as it fell to the ground. Jacob Bjerknes used a collage of these secret images to analyze the structure of the extratropical cyclones he had visualized in Norway thirty years before with a few score barometers and a couple of colored pencils. The advance was striking and compelling. Already in the early 1950s, Harry Wexler, head of research for the Weather Bureau, was planning for a space-based storm warning network. By August 1959, rocketry and television technology advanced enough to allow the first cloud picture transmission from space, using cameras aboard the Army's Explorer VI. The next Explorer carried the first meteorological instrument designed for space by Verner Suomi and Robert Parent of the University of Wisconsin. It monitored the heat emitted from the top of the atmosphere. Images of heat (infrared emissions), rather than visible light, are crucial for monitoring the motion of storms at night. Cold cloud tops—representing the tallest thunderstorm towers—contrast vividly with the warm surface of the Earth.

Satellites were as much an instant success in meteorology as radar. In 1961, yet another hurricane threatened the Texas coast. This time Weather Bureau forecasters caught it early on a crude television picture from space, taken by an experimental weather satellite launched that spring. With the ample warning time, authorities convinced more than 350,000 residents of the Gulf Coast to leave before Hurricane Carla made landfall less than 100 miles west of Galveston. The largest evacuation in the country up to that time held the death toll to 46, most of them people who had refused to heed the warnings.

Not until 1966 were meteorologists able to see the entire disk of the Earth at once the way GOES-8 and its sister, GOES-9 (over the West Coast), can. The new, global snapshots were taken by a special spinning camera that Suomi and others quickly developed as a late addition to a communications satellite. Meteorologists were delighted by these images, which finally made continuous hurricane reconnaissance possible. In 1974, NASA sent one of these cameras up on a geostationary satellite dedicated to meteorology. Not even a few disappointing episodes of equipment failure and budget delays have interrupted the watch of the Atlantic since then. The two current American geostationary satellites complement European, Japanese, and Russian satellites in a global network that is typical of the international cooperation meteorologists require.

While satellites kept their 'round-the-clock vigil, Air Force Reserve and NOAA crews flew hurricane hunting planes right into the eye of Georges to get precise fixes on the storm's progress and strength. Such flights have been uncommonly successful since their inception, made on a whim by Army Air Force Colonel Joseph Duckworth in 1943. Duckworth was a former Eastern Airlines pilot then stationed at the military's Bryan, Texas, training base. At breakfast on July 27 he heard that a hurricane was approaching Galveston. Duckworth decided this would be an ideal test of his expert instrument-flying abilities through near white-out conditions. So he grabbed a navigator, Lieutenant Ralph O'Hair, and decided to do the unthinkable—fly right into the hurricane. As bad as the surface winds of a hurricane are, the peak velocities in the storm are actually about a third faster a few thousand feet higher up. The tremendous gusts at the surface are representative of these higher winds, which get briefly carried to the surface in violent eddies. Duckworth had no idea of the shear between clouds and around the eye, but apparently he had all he needed—a single reliable engine on his AT-6, a two-seat trainer. Duckworth guided the plane into the storm just after it reached land. Heavy static rendered the plane's radio useless as they dared the updrafts between 4,000 and 9,000 feet. The plane was "tossed about like a stick in a dog's mouth," Duckworth later said, yet O'Hair was able to angle the plane across the winds and into the clearing of the eye. Perhaps it was the times, with a war raging and young men and women full of bravado, but when Duckworth landed, flight weather officer Lieutenant William Jones-Burdick was so disappointed he'd missed the flight that he made Duckworth turn around and take him into the storm. Later they found out that others at the base in Bryan had also flown into the storm that day without even knowing about Duckworth's pioneering flight.

The hurricane hunter flights soon became a regular part of Air Force and Navy weather reconnaissance, though not soon enough to save Admiral "Bull" Halsey from losing a few destroyers in the disastrous 1944 typhoon near Okinawa. Flying into a 1945 hurricane in a B-25 bomber, navigator James Dalton felt the plane "suddenly wrench itself free, throw itself into a vertical bank and head straight for the steaming white sea below. An instant later it was on the other wing, this time climbing with its nose down at an ungodly speed.... I stood on my hands as much as on my feet." In the same hurricane, flight engineer Don Smith noted that the rain made it impossible to see the plane's engines from the window. The water leaked through the seams in the fuselage skin. "I have never seen the water pour in and spurt so before," said the flight weather officer, George Gray. "There was a regular fountain about six inches high that flooded the whole area." After flying in a September 1944 hurricane that drowned hundreds of sailors convoying across the Atlantic, one Army crew noticed that the rain and wind had sheared off 150 rivets. The Navy pilots took dangerous low-altitude approaches into the storm in an effort to monitor sea surface conditions. Not surprisingly, one B-29 never returned from a typhoon near Guam in 1952, and a couple of smaller flying boats disappeared into another typhoon a few years later.

Despite the dangers, scientists were quick to tag along for the ride. Robert Simpson, later head of the National Hurricane Center in the 1970s, was one of the first. Simpson had already seen the horrors of hurricanes as a boy growing up in Corpus Christi, Texas. In 1919 a storm surge drowned 200 people there. The Simpson family fled their house just in time to wade out of the rising water. In 1947 Simpson climbed aboard a B-29 hurricane hunter in Bermuda as part of a joint Weather Bureau–Air Force hurricane mission. The plane headed for a hurricane more than 600 miles east of Miami, crossing about 150 miles of clear, sunny outflow region around the storm. After flying under the cirrus canopy of the storm for about 10 minutes, the crew began to spiral upward, attempting to top the clouds. Simpson had thought it would be possible at about 30,000 feet, but this turned out to be wrong. At 36,000 feet, straining the ceiling of the plane, the pilot quit the climb with cirrus still far above.

At that altitude, however, Simpson saw first hand that the outflow circulation predicted by Ferrel and studied by Viñes was not necessarily a simple clockwise spiral in northern-hemisphere storms. In fact, winds were speeding counterclockwise at 130 m.p.h., similar to the *surface* circulation in the hurricane. The outflow was also more concentrated than expected; Simpson found a (relatively) narrow and warm plume of air that pushed ahead of the storm. "Shafts of supercooled water ... rose vertically and passed out of sight overhead," Simpson said. The presence of water so high was startling and very dangerous. The encounter with water so high and cold (−31 degrees Fahrenheit) came only months after rainmaker Vincent Schaefer discovered that water could cool without freezing to nearly −39 degrees in the laboratory. Each pass of the B-29 "seeded" this high-altitude water and left the wings coated with six inches of ice until the plane nearly stalled. The pilot quickly glided the plane down straight ahead to melt the ice. "The plane was simply too loaded with ice and too near stall-out to risk the slightest banking action," Simpson said.

A few days later the crew was back in the hurricane. After entering the cloudy spiral at 10,000 feet, they took advantage of a recent radar discovery, made during World War II, that showed that the storm inflow organized into parallel bands, explaining why Viñes noted the highly variable rainfall in hurricanes. Spiraling cloud bands form more than 100 miles away and march into the eye single-file. In a band, the storm towers don't compete for energy: they cooperate. Once a few clouds get going, they beckon surface air to converge under them. This clears out the air between bands and enhances the updrafts in the existing clouds. Some small storms may have only one band, but other, large typhoons can have half a dozen. With radar guidance, the B-29 turned into "a canyon in the clouds with huge convective walls ... rising to either side and disappearing into the cloud sheet overhead," Simpson said. To stay under the smooth clouds above the trench, the B-29 had to continually descend as it neared the eye. The ride remained smooth even when the gap between bands closed up with cloud. The crew simply followed a path between heavy radar echoes. "To crew members who on previous occasions had experienced severe turbulence ... while trying to 'plow through' directly to the storm center ... this absence of turbulence was incredible," Simpson reported in *Weatherwise* magazine. The flight had its perils nonetheless. After climbing back to 36,000 feet, the plane lost one engine and then another in thick bursts of smoke. With only two engines left, the heavily laden plane practically glided 400 miles to safety in Florida.

Today, hurricane researchers routinely fly into big storms. Following Simpson's example, some NOAA scientists have flown through more than 200 hurricane and typhoon eyes. The flights not only fill the research journals with new questions but also provide the only means of truly accurate wind and pressure measurements. They are certainly far more useful than the Navy research in the '40s in which scientists tried to locate hurricanes with seismographs. Precision is critical because a storm position error can ruin the attempt by a sophisticated computer model to predict a landfall site. Scientists now drop instrument packets into hurricanes, refining the data input into computer models. Part of the reason this is possible is NOAA's new, high-flying Gulfstream jet aircraft that became operational in 1997. The plane has a high enough ceiling to gather information from the critical steering currents affecting the storm's movements. While satellites can detect the motion of invisible water vapor at these altitudes, and thus track the winds, the direct measurements from airplanes still prove useful because of their great accuracy.

The NOAA forecast of Georges' track through the Florida Keys proved accurate, and the storm caused $250 million in damage as its 100-m.p.h. winds destroyed 180 homes and damaged perhaps 10 times that many. The waves ripped tons of sea grass and manatee grass from shallow water beds and piled it on the fine sands of beaches on the mainland coast. Dead grasses choked the water between the mainland and the Keys as well. During the noontime passage over Key West, the water surged seven feet and the spray and rain and cloud produced a disorienting whiteout similar to a North Dakota blizzard. The rising water lifted boats out of the canals of Little Torch Key, and fish swam in the streets.

The experience couldn't compare to the Labor Day disaster, in part because of the warning and preparation but also because the storm lacked the punch of the earlier catastrophic blow. Predictions of six-foot surges were pretty close to the mark, because Georges had little time to recover from its crossing of Cuba before it hit the Keys. But other hurricanes have shown much speedier intensification. In a way, the weak landfall in the Keys in 1998 is as inexplicable as the pounding Miami took in 1992, during Hurricane Andrew. The growth of these storms is at times so sudden that forecasters can easily get caught off guard. Some meteorologists consider the prediction of hurricane strength to be one of the trickiest challenges in their science. It's hard enough to predict what environment hurricanes will encounter; in addition, hurricanes cope with their environment in seemingly unpredictable ways, sometimes waxing, sometimes waning.

One might suspect that the tropical atmosphere is completely bland, that hurricanes thrive there like blanket-swathed babies fed warmed air right out of the maritime bottle. If this were true, tropical clouds would never grow up to be menacing hurricanes. In the 1930s, Gordon Dunn, a longtime hurricane forecaster for the Weather Bureau, noted that rainy, cloudy air hitched a ride eastward across the Atlantic with small fluctuations in surface pressure, even though surface winds appeared to be relatively unaffected. Herbert Riehl of the University of Chicago spent part of World War II at a special tropical meteorological research institute in Puerto Rico. There he had a good opportunity to measure the tropics at work. Despite nearly unvarying conditions at the surface, Riehl found that somehow, once or twice a month, the atmosphere managed to produce a shower accompanied by one of Dunn's subtle pressure waves. Riehl also found that upper-air shifts initiated Dunn's pressure waves. These unexpected ripples in the upper flow—waves in the easterlies—cause just the right amount of convergence and divergence of air to stir up tropical trouble.

Every three to four days during hurricane season an upper-level easterly wave sets out over the Atlantic from Africa. One in seven transforms into a hurricane on its journey. These small stormy areas usually embark from the semiarid regions between 10 and 17 degrees north. At first they drift at about 10 m.p.h. for a couple of days, slogging through the air cooled from below by upwelling coastal waters and compressed from above by sinking Saharan air. The wave in the winds survives, however, remaining most active at three or four miles' altitude. It is a traveling fault line cracking up the temperature inversion that normally caps trade wind cumulus at around 10,000 feet. The inversion rises behind the (roughly north–south) axis of the wave. The wave launches moisture en masse like a swarm of balloons along the parade route across the Atlantic. Runaway cumulonimbus growth results.

That not all easterly waves form hurricanes suggests the importance of other factors in making hurricanes—including sea surface temperatures, upper winds, and temporary zones of shear. The number of easterly waves varies little from year to year. The number of hurricanes, on the other hand, varies considerably. In 1995, 19 tropical storms roamed the Atlantic basin (including the Caribbean and Gulf); of these storms, 11 were strong enough to earn hurricane status. Only 1933 had more tropical storms (21), and only 1969 had more hurricanes (12).

It was a banner year for William Gray, the Colorado State University meteorologist who began predicting seasonal variations in hurricanes in the 1980s. Gray makes his forecasts based on numerous factors, among them rainfall in western Africa, in the Sahel (south of the Sahara), and along the Gulf of Guinea. A long Sahel drought perhaps inhibited hurricane activity in the 1970s and 1980s, when very few major hurricanes developed. In addition, Gray looks above the troposphere to the lower stratosphere, where winds shift every year to year and a half at an altitude of about 70,000 feet. These winds were in a healthy westerly phase in the busy 1995 season, while surface pressures over the Caribbean hit their lowest averages in 50 years, another sign of frequent hurricanes. Gray also says that El Niño, the warming of the tropical Pacific, coincides with weak hurricane seasons in the Atlantic. When El Niño raged in 1991–94, each Atlantic season had four or fewer hurricanes, consistently below the average of six. In 1994, not a single hurricane formed during the peak of the season (August through October).

El Niño may influence Atlantic hurricane frequency in ways as yet unclear to scientists. Gray is especially interested in tracking down the influence of oceanic circulation on the frequency of hurricanes. Fluctuations in the cool, salty water that sinks near the Arctic and crawls along the bottom of the eastern Atlantic may have significant effects. Meanwhile, El Niño's most obvious possible influence is in the circulation of the atmosphere. The equatorial warming builds huge areas of thunderstorms over the Pacific that tamper with midlatitude westerlies and also tend to push away the trade winds. This blockage in the tropical circulation takes some of the harmony out of the wind profile. In 1995, without an El Niño, winds 40,000 feet over the tropical Atlantic did nothing to inhibit uniform easterly flow.

Easterly waves have an intimate relationship with the upper-air flow near them. If the clouds in the wave connect with a migrating area of diverging air at the top of the troposphere, they get sucked up even faster, pressure starts falling rapidly at the surface, and winds begin to spiral inward with speeds approaching 40 m.p.h. The thunderstorms begin to coalesce into a band or two and wrap in toward the low-pressure area.

Now the wave is considered a tropical depression or storm with about a 50-50 chance of making it to the big time–hurricane strength. It helps if the thunderstorms can pool their warm air around themselves at higher and higher altitudes, just like the cooperating thunderstorms of mesoscale convective complexes. This isn't easy. At high altitudes, an easterly wave actually begins life a few degrees cooler than its environment, because this air is mostly from outside the storms, not from the sea surface below them. The wave probably throws its temperature switch only if the cloud towers are stout enough to withstand the inevitable entrainment of energy-less outside air into the stormy fuel tank. Pooling upper-level warmth can reach an excess of 10 to 15 degrees Fahrenheit in typhoons, enabling the storm towers to invigorate the most essential part of the tropical cyclone engine—upper outflow. Sometimes growing tropical storms get a power boost here simply by traveling under the lingering upper-air divergence left behind by a previous hurricane. The new storm tries on the old upper-level warm core, and often it fits perfectly, outfitting the new engine with a track-tested turbocharger. The growing storm's spreading anvils become menacingly flared pipes spewing the hot exhaust of a souped-up speed machine.

Even with a good exhaust system, however, the hurricane can choke on its own latent heat release. The clouds warm the air so much that eventually new moisture is unable to condense. Colder air ventilating through the hurricane from outside the clouds is the condenser that completes the heat engine design of the hurricane.

To really tune up the engine to a feverish hum, the center transforms into a ring of storms—the eye of the hurricane. Here low pressure draws air down from the tops of the cloud towers. The sinking air warms and evaporates droplets, clearing away stray clouds. Circling in a plane in the eye of Typhoon Marge for more than four hours in 1952, Robert Simpson noted air 14 degrees warmer at 8,000 feet and 32 degrees warmer at 18,000 feet than in the surrounding cloud bands. "This astonishing temperature gradient indicated that the storm had as great a concentration of potential energy as has ever been detected in the atmosphere," he wrote at the time.

Nor is the eye itself just any empty space in the storm. Some perfectly good eyes are filled with scruffy cloud. In Typhoon Marge, Simpson found that the air spilling into the eye from the top collected into a cloudy cone about 8,000 feet high at the center, sort of the way tea leaves gather at the center of a vigorously stirred cup of hot water. Clear or not, the eye opens when spiraling winds contract and accelerate into a radius where they are so fast that the centrifugal effect flinging them outward is strong enough to balance the suction of the inner low pressure. (A nascent hurricane earns a name as a "tropical storm" when it develops 39-m.p.h. winds; only when it reaches minimum hurricane strength of 74 m.p.h. does the eye form.) The balance has to be perfect. Many times a cloud band of a tropical cyclone will encircle a space, and the outflows above will leap into the eyelike abyss and clear out the clouds. Yet the space won't hold and the storm will refuse to build toward hurricane strength.

Concentration of energy, not sprawling size, makes the hurricane a force to reckon with. The average hurricane spreads its 74-m.p.h. or greater winds across a circle only about 150 miles wide, though tropical-storm-force (39-m.p.h.) winds extend considerably farther. Only the eyewall and immediately surrounding band are likely to climb to the top of the storm. In the eyewall the updrafts don't necessarily reach the supercell strength seen over tornadoes. Nonetheless, air might ascend faster than 40 m.p.h. in the eyewall and nowhere else in the hurricane. This towering fortress of updrafts is the hub of the storm, the connection between inflow and outflow.

The importance of a smoothly running eyewall explains why hurricanes can't abide much vertical wind shear—the layered opposition, crossing, and acceleration of winds that El Niño encourages in the tropics and that supercell thunderstorms eat up in the Plains. With too much shear, the eyewall updrafts wouldn't remain upright enough to hook up the Earth's surface with the stratosphere. Poleward from about 35 degrees latitude wind shear is a daily fact of life, and therefore so are layered stratus, not cumulonimbus. In the tropics, however, wind blows steadily from the east throughout the troposphere. When an intruding upper-level trough occasionally bends high-level westerly winds too far Equatorward, the resulting shear can destroy otherwise smoothly running hurricanes. This is one reason why, in November, the last month of the official Atlantic hurricane season, few hurricanes can form. In addition to cooling sea waters, the midlatitude jet stream begins to migrate south toward its winter home, ripping apart nascent tropical storms, especially in the Caribbean.

Scientists learned much of what they know about hurricane intensification by trying to diminish the storms with cloud seeding. In those experiments, performed in the 1960s and '70s in the government's Project Stormfury, the eyewall dynamics figured prominently. Without the potent concentration of resources by the hurricane in the eyewall, the seeding plan would have had no hope. The scientists tried to seed the band just outside the eyewall. Enhancing these clouds might release enough heat, they thought, to encourage a new eyewall outside the natural one, "short-circuiting" the old eyewall. Calculations showed the storm would weaken by up to 30 percent if seeding could double the eye diameter.

Unfortunately for the seeders, few major storms developed in the Atlantic basin; a 20-year lull in hurricane activity set in. In 1961 Hurricane Esther offered a rare opportunity. The seeders saw a promising 10 percent reduction in winds from a small amount of silver iodide, but the next day's follow-up drop missed the target area. In 1963, winds dropped in Hurricane Beulah two days after the seeding began, probably because of an errant silver iodide drop again. Beulah went on to create a record 115 tornadoes in Texas after landfall, anyway. Hurricane Debbie of 1969 weakened on cue after the hurricane seeders dumped silver iodide near the eye five times over 10 hours. Within six hours of the last dump, the eye lost more than 30 m.p.h. and became a weak hurricane as it widened. With a respite from seeding, the hurricane was back up to full strength within a day. Another intensive seeding brought winds down to below 100 m.p.h. again.

Stormfury died a political death as well as a scientific one. No one knew for sure whether the seeding influenced where the storm would strike. The scientists seeded storms that were far out over the ocean and seemed unlikely to strike land, but neighboring countries complained that the seeding might reduce the rains the hurricanes brought to otherwise dry islands. In 1983, hurricane seeding experiments ended.

Ironically, the hurricane modification attempts revealed to researchers how some hurricanes intensify and weaken naturally. Critics of the project insisted that storm seeders should try more "control" experiments. They wanted to see if the hurricanes would weaken even if the seeders dumped placebo into clouds. Some storms naturally fizzled with or without seeding—especially storms with large eyes or diffuse structure. If anything, the seeders' strategy focused on strong storms with small eyes. Some meteorologists began to suspect that either a storm is likely to weaken naturally or the seeding only works when the eye is small.

In the early '70s meteorologist Hugh Willoughby was flying through typhoons out of Guam. He had visited the Stormfury scientists in Miami,

Before and after the hurricane: South Carolina's Folly Beach suffered Hugo's wrath in 1989. NOAA photos.

so he knew about their eyewall replacement seeding. He urged meteorologists in his Navy squadron to look for naturally occurring eye widening in the Pacific; amazingly, the crews found the process again and again with radar. Many storms temporarily developed concentric eyewalls, with the outer wall eventually replacing the original inner wall. Then, working for NOAA in 1979, Willoughby saw a similar pattern while flying in Hurricane David. The storm developed concentric eyewalls; then the inner wall, choked off from the storm circulation, died off. The eye temporarily expanded and the storm lost strength as the outer eyewall became preeminent.

Some of the seeders became suspicious as they looked over the results of their experiments. Some scientists thought Hurricane Allen in 1980 had been seeded, because the eyewall had expanded and the inner pressure had weakened significantly before a new tight eyewall closed in around the center. But in fact no one had dropped any silver iodide into the storm. "It was exactly the case that made us think it would be impossible to disentangle the natural behavior from the expected effects of seeding," says Willoughby, now director of NOAA's hurricane research division in Miami. One Air Force reconnaissance crew in Allen reported a small clear eye surrounded by 130-m.p.h. winds. But during the next flight into the storm, the NOAA P-3 encountered nothing but cloud cover within the eye, which had widened to 40 miles across. The storm was in the middle of an eyewall replacement.

Willoughby realized that the natural eyewall replacement might explain fluctuations in storm intensity. At first reviewers criticized the hypothesis as "unscientific drivel," he says. But now researchers have seen the eyewall replacement cycle many times. The belated recognition of these natural events demonstrates the complexity of most hurricane radar signatures. "If I hadn't believed it, I wouldn't have seen it," Willoughby says. He thinks that the shear along the eyewall might get "spun out and filamented into a ring" to create the new outer eyewall. Sometimes, however, the new rings of intense spin in the storm merely replace similar features, Willoughby says. The replacement process might be the most visible effect of a more subtle redistribution of spin constantly played out in storms.

Another way of looking at redistribution of spin in hurricanes is to compare it to the Rossby waves of the jet stream. These waves are modulations of flow that help the atmosphere conserve spin in the upper air—spin that originates in part from the Earth's rotation. The air flowing north or south encounters a new environment with different spin and compensates by turning back and forth across latitude lines. In the hurricane, according to recent modeling studies by Michael Montgomery at Colorado State University, such Rossby waves develop from the asymmetries in convective towers popping up along the inner rain bands. These bumps in the flow jar the air in and out relative to the eye, setting up waves as the winds adjust to the different spin intensities at different distances from the eye. The resulting Rossby-type wave fluctuations may explain cycles of eyewall expansion and contraction.

Modeling eyewall replacement on computer has so far been difficult. And without understanding exactly when eyewall replacement will occur, forecasters are still left with a thorny dilemma as a hurricane approaches landfall. The National Hurricane Center tries to issue storm intensity predictions using the Saffir–Simpson scale, developed by South Florida engineer Herbert Saffir with Robert Simpson in the early 1970s. The scale relates likely winds and storm surges to the eye pressure. There's a big difference between a Category 1 storm, with 74- to 95-m.p.h. winds and up to a five foot storm surge, and a Category 5 storm, with sustained 155-m.p.h. winds and maximum storm surges exceeding 18 feet. As helpful as the scale is to local emergency managers who plan coastal evacuations, the Hurricane Center often sees errors of at least one category in its 24-hour advance landfall predictions.

No storm presented a thornier intensity problem than Andrew in 1992, one of the four strongest hurricanes to strike the United States this century. Andrew fired up its engine as an easterly wave from Africa. By the time the wave reached the Cape Verde Islands it found gentle, warm waters conducive to thunderstorms. The stormy area officially became a "tropical depression" on August 16, two days into its Atlantic journey. With uniform winds up to 50,000 feet, the low-pressure area continued organizing, reaching tropical-storm strength on August 17. A low-pressure trough briefly threatened Andrew three days later, but it quickly split and moved out of the way. Instead of causing damaging shear, the high-level southeasterlies brushed the edge of Andrew's flat-top anvil, probably helping the young tropical storm pump out its hot breath of moisture. In part the storm may have been saved by its diminutive profile. With gale-force winds only 100 miles across at its peak, Andrew was small but incredibly feisty.

Less than 48 hours after the tropical storm rubbed elbows with the midlatitude trough, the sun rose on the season's first hurricane. In fact, after two quiet seasons, the storm was the first African wave to develop in two years. Andrew began to steam due west toward Miami, an unlikely path cleared by a long ridge of high pressure to the north stretching off the southeastern United States. Andrew sailed over 86-degree waters in the Atlantic and supercharged itself to category 5 winds. This doubling of winds took only a day and a half, a transformation whose rapidity marks the worst hurricanes. For example, forecasters watched in amazement in 1989 as Hurricane Gilbert's central pressure plummeted more than two inches in a day, to 26.22, over the Gulf of Mexico. With 185-m.p.h. sustained winds, it was the strongest hurricane (and lowest sea-level pressure) ever recorded in the western hemisphere. The storm killed 318 people, more than 200 of them in Mexico after looking for a while like it might hit south

Texas. In 1997, Hurricane Linda fooled forecasters with a similar rapid intensification: when its winds first reached 39 m.p.h., forecasters figured Linda would barely make hurricane intensity in three days. Instead, its eyewall winds roared to 185 m.p.h. in that time. For a time, forecasters began to worry that Linda might actually become the first tropical storm to strike southern California since 1939. A September storm that year killed 45 people when more than five inches of rain a day caused flooding in the Los Angeles and San Diego areas. Linda took advantage of unusually warm waters off the Baja coast—a result of the record El Niño—to climb northward. But the storm fell apart before remnants dumped rains on the United States.

As Hurricane Andrew spun faster and faster, it squinted its eye to only five miles across. But then the storm hit the Bahamas. Andrew kicked up a remarkable 23-foot-high storm surge. But with only five feet of water at low tide, the Great Bahamas Bank offered so little heat energy for its voracious guest that Andrew began to sputter. Three hours over the shallows was enough to slow winds by 40 m.p.h. It was a do-or-die moment for Andrew. Its response was to throw a new wall of cloud up to the sky, cutting off the old eyewall from inflow winds. As the small eye withered away, the new eyewall desperately sucked up heat and moisture from the Gulf Stream off Florida. The last-gasp effort succeeded. The outer eyewall quickly tightened, and winds picked up again to 140 m.p.h. before sunrise on August 24.

The infusion of warmth from the Gulf Stream shows how critical water temperature is to a hurricane. No hurricane can survive for long above the cold waters outside the tropics. Similarly, slow-moving hurricanes easily exhaust the warmth beneath them. Usually, a storm can merely skim the cream off the milk: the upper 100 feet or so of ocean is a warm, buoyant layer floating atop denser, colder deep waters. A hurricane usually only drinks from the tepid top of the sea. But hurricane winds also stir the surface and draw up the cold waters with turbulence. If a tropical cyclone lingers, it gets a chilly draft from this upwelling; similarly, if it crosses the upwelling wake behind another cyclone, it will sputter.

Hurricane forecasters also, then, have to consider the vagaries of the sea surface. Over the Gulf Stream, for instance, where the warm surface layer is an extra couple hundred feet deep, a lazy hurricane can intensify with no trouble. Sometimes these warm ocean currents develop kinks that break off as separate eddies. Hurricane Opal in 1995 found one in the Gulf of Mexico and quickly intensified before landfall in Florida. At the same time, Opal met the tail end of an upper-level trough moving off the Texas coast. The trough brought strong winds beside Opal's outflow—but not too close for storm-destroying shear. In a move as delicate as when trapeze artists grasp hands high above the nets, the two aerial disturbances over the Gulf enhanced one another briefly, helping lower Opal's central pressure. The relative importance to Opal of the watery warming below and the enhanced outflow above is one of the trickiest questions in hurricane intensity research.

Andrew's response to the Gulf Stream, on the other hand, was unambiguous. At the National Hurricane Center, then in Coral Gables, meteorologists watched the eyewall 100 miles away on radar as it grew and then shrank. Temporary hopes that the storm might fizzle before landfall were dashed. Suddenly the forecasters found themselves within the very winds they were predicting. Director Robert Sheets had already sent two of his forecasters to Washington, D.C., to continue operations in case the storm knocked out power and communications. But backup generators kept the meteorologists in the storm hooked up throughout the night.

The building itself began to shake as winds whipped to 115 m.p.h. over the roof. An hour later, at 4:45 p.m., the winds toppled the radome from its rooftop mount. The forecasters switched to images from Tampa and West Palm Beach. Just after 5 a.m., Andrew heaved nearly 17 feet of water onto the Bay of Biscayne, narrowly missing the vulnerable condominiums in Miami Beach. At the National Hurricane Center, in the intense winds north of the eye, a 164-m.p.h. gust ripped away the anemometer for good. In the parking lot below, the storm tossed cars around like toys, and later 10 of the Weather Service workers found their houses badly damaged by the winds. A photographer who rode out the storm in a parking garage described Andrew's winds as "the scream of the devil" that rocked the steel reinforced structure. (Similarly, the roar of Hurricane Camille's winds in 1969 registered 120 decibels on one meter—equal to a jet airplane taking off overhead.)

Andrew's eyewall contractions caught Dade County at just the wrong moment. Only seven inches of rain fell, and the storm surge was relatively benign in the city. But the winds seemed unprecedented. Somehow, the (category 4) 145-m.p.h. peak winds caused a record $25 billion in damage. Homestead Air Force Base was a near total loss. So was a nearby trailer park. In all, Andrew destroyed close to 25,000 homes and damaged another 100,000. Oddly, the disaster seemed magnified on the south side of the eyewall, where the forward movement of the storm subtracted 18 m.p.h. from the winds. Many supposedly well-built homes in south Dade County suffered damage more appropriate from tornadoes.

In part, Dade County suffered from its explosive growth. Poor enforcement of construction codes, generally acknowledged to be among the nation's strictest, contributed to the disaster. But meteorologists found other disturbing problems. The radar scanned Andrew over Florida at heights ranging from 1,500 to 25,000 feet. Seven convective cells spun up in the northern eyewall in 45 minutes as the central cloud ring straddled its beachhead. Each intense updraft cell then circled around the eyewall to the southern edge, where it pounded Homestead. Under the embedded thunderstorm vortices, which lasted only about 10–15 minutes each, rain fell 50 percent harder. Afterward, wind expert Ted Fujita surveyed fleeting damage swaths across communities like Country Walk Estates, where some houses collapsed beside others that stood nearly unscathed. Andrew also destroyed a few concrete structures built by Florida Light and Power that were engineered for 190-m.p.h. winds. Fujita concluded that the locally heavy rains may have created microbursts, while shear also spun up fleeting vortices with 212-m.p.h. winds. Fujita identified 30 of these whirls—some only 50 feet wide—in addition to the 14 tornadoes spawned by the storm.

Researchers encountered a similar phenomenon in Typhoon Ida in 1958. Flying in a U2 spy plane thousands of feet above the eye over the Pacific, Air Force pilots had photographed knots in the eyewall where the clouds twisted around a small, embedded vortex. Then in 1989, Hurricane Hugo gave NOAA researchers the ride of their lives. A small, 65-m.p.h. vortex within the 155-m.p.h. eyewall rocked the plane unexpectedly. The crew had to shut down an engine that spun out of control. For several harrowing hours, the researchers circled inside the eye, burning off weighty fuel before the pilot dared attempt to take the injured plane home through the storm. Meanwhile they saw the vortex circle the eyewall every 10 minutes, spinning off thunderstorm cells.

"We almost died in Hugo, and when the same thing happened in Andrew we got to thinking, 'Man, the thing's not only out to get us but it's out to get our houses,'" said Willoughby, whose roof was damaged in Andrew.

Ida, Hugo, and Andrew all were intensifying storms when the vortices formed in their eyewalls. These small whirls embedded in the storm's circulation appear to be deformities in the elegant natural symmetry of the hurricane. This symmetry more and more appears to be an illusion. None

Hurricane Emilia gazes up at space shuttle astronauts over the eastern Pacific in July 1994. With winds exceeding 150 m.p.h. during this stage, the eye suggests complexities scientists can't yet explain, such as the inconsistent slope and narrow spiraling rolls along the inside edge of the eyewall. NASA Photo courtesy of Gregory J. Byrne.

of the eyewall's spasms and contortions alters the basic picture—that a hurricane is a remarkably steady-state vortex once it gets going. But for years observations have suggested that in fact hurricanes have complex, ever-changing shapes. As early as 1946, Robert Simpson reported spots of anomalous pressure rotating around the eyewall, and of course his flights in 1947 showed a tongue of warm outflow ahead of the storm. In the 1950s regular aircraft observations began to show that hurricanes wax and wane, rather than maintain steady strength. The aircraft observations also showed that hurricanes stagger in toward landfall like drunks unable to walk the line: they zigzag back and forth. By the end of the 1960s, mathematical models showed that this drunken sway is an oscillation caused not by steering currents but instead by the internal wobble of the hurricane structure.

For many years the best technique forecasters have had for hurricane intensity growth was a detailed historical review of satellite imagery made by NOAA meteorologist Vernon Dvorak in the 1970s. The method is a pictorial typology of hurricane development that gives meteorologists a norm of behavior. They use the Dvorak method to locate the eye in half-hidden rain bands, and then project the next stage in the storm's development. But the evidence clearly shows that there's little regularity to such development on a day-to-day basis.

The first mathematical models of hurricanes, in the 1960s, also suggested the inadequacy of a simple depiction of hurricane development by their obvious failure to make realistic hurricane intensification. The models were able to make hurricanes grow up from small disturbances quite well. But they made every disturbance—not one in seven—grow to a full-blown storm. Major advances in such modeling came in 1969, with work by Katsuyuki Ooyama and Richard Anthes. Anthes, then a graduate student at Pennsylvania State University, developed a 3D model of a hurricane that was able to show the asymmetries of storm structure.

As Georges left the Keys and headed north in the Gulf, it intensified as the computer models indicated. Now hurricane forecasters had to deal with another crisis. Nowhere in the United States is an accurate track forecast more critical than along the Louisiana shoreline. The recurring nightmare of meteorology is that a storm surge will hit New Orleans. Here a city of more than a half million people lies several feet below sea level, protected in part by levees, but with an underground pumping system admittedly inadequate for the task posed by a major hurricane. Worse yet, few roads lead out of the city through the surrounding bayou. A direct hit here could easily make Hurricane Andrew's destruction look very small. As Georges approached, the mayor of New Orleans told his constituents to board up their homes and leave, and most did, flooding motels as far north as Tennessee. In all, well over a million people evacuated the Gulf Coast.

Better computer modeling and intensive satellite and aircraft reconnaissance have increased accuracy of hurricane track forecasts by close to 1 percent a year. The storm is likely to strike somewhere within 100 miles of projected landfall. But at the same time, Americans are migrating to the coasts. So, even as meteorologists can focus their warnings on smaller and smaller stretches of the shoreline, more and more people have to evacuate. The exodus costs close to $600,000 per mile of evacuated coastline.

The uncertainties of hurricane forecasting ensure that inevitably some of the evacuations are false alarms. The Bermuda high and similar upper-air features over the Pacific dictate the eventual poleward turn of tropical cyclones. But the stronger the storm, the more control it has over this recurvature. Basically, the high pressure and the low pressure move around a common center of gravity. Usually, the spreading high is so much bigger that it bullies the compact, converging hurricane into making the turn. A large storm has more bulk and consequently throws its own weight around, forcing the high to accommodate it. This is especially true in the Pacific, where supertyphoons have all the time and heat in the world to grow. They hook up to the high beside them and pull it along like a motorcycle and its sidecar.

Intensifying storms also make their own track, to an extent. Tiny Hurricane Camille of 1969 became extremely intense in a hurry as it headed north over the Gulf. To accommodate the sudden increase in exhaust, the upper circulation of the storm concentrated into a jet of air billowing out the northeastern side of the hurricane. This enhanced the high-pressure area to the right of the storm. In response, Camille turned toward the northwest. On the other hand, weakening storms are notoriously erratic. They respond sensitively to their environment. Hurricane Gordon in 1994 took a bizarre S-shaped path from Central America to eastern Cuba and back to the Gulf before crossing the Florida peninsula on a northeasterly track. It accelerated out over the Atlantic in what seemed to be its final throes of recurvature. But then it hit a trough to the west and weakened. Gordon made an about-face and looped back toward Florida and South Carolina.

At the National Hurricane Center in 1998, forecasters watched Georges strengthen to close to 100-m.p.h. winds again over the Gulf. Fortunately, it didn't do anything fancy. Just after 11 p.m. on September 27, Georges finally made landfall for the last time, not far from Biloxi, Mississippi. The surge was only about seven feet, enough to pound numerous beachfront houses into submission, but a far smaller threat than the worst storms in the region's history. Seawater was waist high in downtown Mobile, however, and high waves destroyed fishing huts along Lake Pontchartrain. Of the four people who died that night, two were killed when they accidentally set their homes on fire during the power outage. New Orleans, fortunately, suffered limited wind damage.

The winds in Georges quickly dropped below hurricane strength early on Monday. If hurricanes are temperamental monsters of the deep, addicted to the hot waters of the tropics, they also are as helpless as fat flippered seals on land. Even if the land below is hot, it lacks the moisture necessary to infuse this heat efficiently into the storm's veins. So hurricanes often crawl up onto land at the last moment to die. The only way a storm can survive crossing onto land is if it moves fast enough to make the journey brief. For instance, Mitch whipped itself into a 180-m.p.h. vortex, then maintained that frenetic pace for an incredible 15 hours (and its Category 5 status for a day and a half). But this stalwart hurricane, probably one of five or six strongest of the century in the region, succumbed to its landfall over Central America. The winds died down to tropical-storm strength on its slow passage inland. (The storm nonetheless reformed back over the Caribbean when it rebounded toward a final landfall over southern Florida.) Andrew was one of the notable exceptions, however, avoiding an early end as it crossed Florida in 1992. The hurricane sped at 20 m.p.h. over the watery Everglades, emerging over the Gulf four hours later as a weakened but undeterred threat to Louisiana. Winds picked up again to 120 m.p.h., and waves topped 65 feet offshore before the eye made its final dash on land west of Morgan City, Louisiana.

Three years later, Hurricane Opal moved fast enough to whip the Atlanta suburbs with 70-m.p.h. gusts. In 1954, Hurricane Hazel held together long enough after speeding into the Carolinas at 60 m.p.h. to reach Toronto. The Long Island Express of 1938 dashed ashore fast enough to flatten millions of trees in Maine and northern Vermont before it exhausted itself. Winds atop Mount Washington reached 186 m.p.h.—remarkable even on that blustery summit.

Georges did not die a quiet death either, but only because it was so slow. It butted heads with a ridge of high pressure over the Southeast and began moving ahead at only 6 m.p.h. Rain quickly became the enemy of Gulf Coast residents. Pensacola received 20 inches of rain in 24 hours, and the Shoal River rose to more than 24 feet—six feet above the old record—in a matter of days. The damage was as extensive in the panhandle as it had been in the Keys.

In 1969, Hurricane Camille was similarly persistent after landfall. Like Georges, it began as a West African wave. Over the Atlantic, it teamed with a leftover warm top from a previous hurricane. On August 15, it had 115-m.p.h. winds and swept across western Cuba. After intensifying over the Gulf, it slammed into Mississippi over Clermont Harbor near midnight on August 17. At Pass Christian, Mississippi, the storm tide rose an incredible 24 feet. The surge was twice as high as that from Hurricane Audrey, which killed 390 people when it struck near Cameron, Louisiana, in the dark of night in 1957. At 26.85 inches of mercury, Camille's landfall pressure was second in U.S. history only to the Labor Day 1935 hurricane. The National Hurricane Center continually posted warnings about the "extremely dangerous storm," and many people evacuated. Nonetheless, at least 144 people died along the Gulf Coast, as nearly every low-lying building along the beach was swept away under the eyewall.

Hurricane Camille quickly weakened over land, but its remnants kept moving north. At first all that tropical moisture over land seemed beneficial. A half inch of rain fell over western Tennessee and Kentucky, where crops had been baking in a serious drought. Two days after landfall, however, the moisture streamed over the mid-Atlantic states while a front approached from the north. The hurricane's remnants squeezed up over the Blue Ridge Mountains and triggered massive thunderstorms over the James, Rivanna, and Maury rivers of Virginia. The tiny rural communities along these mountain streams bore the full brunt of the airborne waters of the Gulf of Mexico. In the dark of night, 153 people died in flash floods. The cruel rains covered an area only a few miles wide. The next morning, residents of nearby communities began to suspect the extent of the disaster only when their coworkers failed to show at their jobs.

Georges did everything a hurricane could do to disrupt the lives of Americans, but by a minor miracle it failed to produce a catastrophe on the scale of the floods in the Caribbean. It treated the Keys relatively kindly, it didn't intensify too much over the Gulf, and it spared New Orleans. Meteorologists practically never get such a dangerous hurricane that behaves so well so many times. Georges was by turns a killer, a survivor, and a relatively gentle giant. It was a long-running act of alternating teases and tantrums—a meteorologically fascinating prelude to the brute force of wind and rain that Mitch later mustered.

This is the irony of the modern age of storm science. Long ago, scientists could dream of the ultimate storm structure. They could imagine the typical storm path. As recently as this century, they recognized the average storm life cycle. These conceptualizations greatly advanced our understanding of blizzards and tornadoes and thunderstorms, not just hurricanes. The ideas behind them were generalizations, however, unworthy of the chaotic atmosphere and its extreme moods.

Now that radar and satellites peer into storms with greater detail and computers simulate them with more realism, meteorologists see through the once-simple theories. They analyze the character of storms with great insight. Meteorologists at one time could confidently limit their research to detailed studies of single weather events, because they figured that understanding one storm meant understanding a whole class of them. Then, hurricanes quickly emerged as a class of storms that defied this treatment. They were clearly the ultimate storms. They earned a reputation for having personalities of their own. They began to live up to the names meteorologists gave them. Someday, it appears, observations will be so good and theories will be so subtle that every storm, not just hurricanes, might earn a name. Perhaps that's what our new perspective on weather is all about.

CHAPTER TEN

An Awesome Chaos

In this last decade of the millennium, a small group of meteorologists has headed each summer to the mountains of the American West in search of truth. These scientists scale great heights far from civilization. They peer into the pitch black night sky to marvel at weird and wonderful lights.

This annual pilgrimage is not to favorite cult locales like Mount Shasta or Roswell, New Mexico. Instead, the scientists go to the Langmuir Laboratory on 11,000-foot Mount Baldy, or to their colleagues' home-built labs atop a ridge in Colorado—places bristling with low-light video and television cameras, special telescopes, and other electronic imaging and listening gear. The meteorologists have a purely scientific question to settle, one that may someday change the way we look at thunderstorms.

A decade ago, most meteorologists would have considered this mass pilgrimage frivolous. Indeed, the questions that prompted it were practically unthinkable. Only a few scientists knew that in the days before cities became perpetually bright, many people reported remarkable shapes and colors of lightning. Later, in darkened cockpits miles above civilization, airplane pilots witnessed blue shafts of light explode up from storm tops. The ever-provocative lightning expert Bernard Vonnegut was one of the few to take these reports seriously. He described them as "lightning to the ionosphere." He postulated that these discharges traced electrified channels of thin air all the way to the heavily charged atmospheric layer above 50 miles in altitude.

Attitudes changed with one freak observation in July 1989. John Winckler, a University of Minnesota physicist, aimed a sensitive new research television camera at the night sky for a test tape. When Winckler looked at the tape later, he noticed a brief explosion of faint light a dozen miles above a thunderstorm 150 miles away. When NASA scientists learned about Winckler's video, they worried that such discharges could threaten the space shuttle. They immediately began shuffling through their own video library to see if the TV camera on the shuttle had recorded similar images from space. Sure enough, it had taped nearly two dozen bolts into the blackness of space. Each leapt toward the stars moments after a flash of lightning.

These black-and-white images of the strange fireworks above the storms whetted scientists' appetite for more. In 1993 and 1994, while others gathered on mountaintops overlooking the grasslands of Colorado and Kansas, University of Alaska scientists Davis Sentman and Eugene Wescott took color video cameras with them on night flights over the Great Plains. From the NASA jet, they found a whole new world of unidentified lights above storms. They recorded close to 1,000 jellyfish-shaped bursts of red light that lasted less than a hundredth of a second each. The body of light, 10s of miles across, reached close to 60 miles high; the tentacles hung down toward the thunderstorm but remained about 10 miles above the clouds. At meteorologist Walter Lyons' laboratory high atop Yucca Ridge in northern Colorado, scientists saw these electrical displays above storms over the horizon. They telephoned their observations to Rhode Island, where MIT scientists used a special antenna for electromagnetic fluctuations to confirm that nearly every burst coincided with a powerful flash of lightning lowering positive charge to the ground.

The University of Alaska–Fairbanks scientists later dubbed the vast red fireworks "sprites." In shape and extent, sprites are nothing like ordinary lightning. They often appear in clusters and spread over 1,000

A sprite fleetingly graces the upper atmosphere many miles above a plains thunderstorm. Image courtesy of Walter A. Lyons/FMA Research.

cubic miles. Atop 14,000-foot Mount Evans, one of the highest points in Colorado, two researchers measured the energy emitted by the sprites and identified the source as excited nitrogen atoms in the sky.

The scientists found much more than the sprites, which are relatively common. They were amazed to discover "blue jets" of light shooting upward from the clouds at 300 times the speed of sound, opening out like searchlights that reached 30 miles high. These seemed to be quite rare and limited to powerful storms; one thundercloud over Arkansas, however, produced more than 50. In one seven-minute span, the same storm launched 30 geysers of blue light that climbed only a few miles from the cloud top. These "blue starters" shot up above shafts of golfball- and baseball-size hail. When Sentman and Wescott hunted for the lights over South America in 1995, they found plenty of red sprites, but no blue jets. That same summer, from Lyons' Yucca Ridge perch, scientists detected curtains of diffuse light—"elves"—more than 100 miles across the ionosphere. The elves shimmered far above thunderstorms after lightning began but before the sprites appeared.

Fanciful names aside, the sprites, blue jets, and elves conjure whole new visions of thunderstorms. To explain the new class of phenomena, some physicists are reformulating an old idea—that cosmic rays cause terrestrial electrical activity. Meteorologists who have been pondering the significance of positively charged lightning have new questions to answer as well. Most importantly, though, the surprise discovery requires a reexamination of the entire global electrical circuit that is continually recharged by thunderstorms. Meteorologists cannot ignore sprites, jets, starters, and elves as they take stock of the role of storms in our world. Many other discoveries are forcing meteorologists to rethink storms as well. Some of the most prominent ideas of this century may not last long in the next.

Part of this rethinking of storms, including sprites, stems from a challenging global perspective thrust upon meteorology. After World War I, meteorologists found a global boundary between cold and warm air and called it the Polar Front. Careful observation showed that disturbances along this front could travel all the way around the world, dropping snow, then rain; blowing dust, then building ocean waves. Satellites eventually showed these phenomena daily. Meteorologists responded by making unanticipated connections between events separated by oceans and continents.

More than any other Americans, Californians have grown accustomed to this global perspective in the 1990s. In the winter of 1995, nearly 70 feet of snow fell in the Sierras. The ski resorts remained open until July.

At lower elevations, rain fell furiously, especially on January 10, when a single storm dropped 10 inches on Sacramento and flooded the area. San Francisco had 26 days of rain that month alone. Houses in the city collapsed into sinkholes.

In another storm in March, Interstate 5, the state's main north–south artery, washed out at Coalinga, and the Russian River swelled 32 feet in one day. Two people drowned and the state endured nearly $500 million in storm damage that winter.

In 1997–98 the rains returned with a vengeance, setting records nearly everywhere in the state. Santa Barbara had an unprecedented 21.74 inches of rain in February. San Francisco had more rain than in any of the preceding 100 years. Some communities along the Big Sur coastline were cut off from the outside world by mudslides along coastal Route 1. Dozens of people died as windy storms lashed the coasts, houses fell into the sea, trees battered roofs, and floods swept away neighborhoods. The damages again reached more than half a billion dollars for the season.

These storms were stirred by El Niño, the planet's biggest frequent climate tantrum. This irregular warming of the coastal waters off Peru, usually around Christmastime, attracted no scientific attention for many years. No one would have linked it to devastating storms in the United States. At the turn of the century, Gilbert Walker, a British meteorologist in India, noted a cycle of equatorial Pacific air pressures that he called the "Southern Oscillation." Pressures rose in the western Pacific when they fell in the central Pacific, and vice versa. In the 1960s, while teaching at UCLA, Jacob Bjerknes made one of his greatest contributions to meteorology by linking the Southern Oscillation to El Niño with a sweeping, trans-Pacific feedback of wind and water. Normally the equatorial Pacific surface is more than 10 degrees Fahrenheit warmer in the west than in the east. The normal easterly trades blow unimpeded across the ocean, attracted by warmth-induced low pressure. But when the pressure shifts, Bjerknes said, the easterlies begin to stall and can no longer push back warm surface waters from the Peruvian coast. Upwelling of cold, deep water ceases there, ruining the fishing industry. Storminess spreads the warm water eastward, and a full-fledged El Niño rages. The anomalously strong and widespread thunderstorms migrate eastward from their usual haunt closer to Indonesia. As a result, these convective towers spread moisture into the jet streams feeding U.S. storms. The strong outflow can also bend the jet, creating anomalous paths for extratropical storms that normally grow over the North Pacific.

Mark Cane and Stephen Zebiak of the Lamont–Doherty Earth Observatory ran the first successful simulation of this El Niño feedback in 1984. Their computer model showed the same erratic two- to seven-year cycle seen in nature. More recently, modelers at NOAA, NCAR, the Max-Planck Institute in Germany, and the Scripps Institution of Oceanography in La Jolla, California, have made headway in linking models of the regional El Niño shifts with models of the atmosphere's circulation. The 1997–98 El Niño showed the potential of this modeling, and successful forecasts helped people prepare for El Niño months in advance. Californians contracted for roof repairs, Cubans harvested their sugar early, and South Africans braced for drought.

But modelers still couldn't simulate the rapid spread and deepening of the record El Niño. Ironically, thunderstorms may have been a reason. Packs of thunderstorms sailed out of the Indian Ocean into the Pacific on a 40- to 50-day cycle discovered in 1971 by NCAR researchers Robert Madden and Paul Julian. The Madden–Julian oscillation, as the stormy surge is called, was unusually effective in 1997, perhaps escalating El Niño toward record warmth by sending pulses of warm water eastward. If so, then successful El Niño computer models may someday link the thunderstorms of the Indian Ocean to the storms of the United States.

Global warming is an even vaster give-and-take in the atmosphere that is changing storm science. Most meteorologists spend little time thinking about such connections, because at present the effects of the warming appear to be subtle, if at all measurable. Tom Karl, a leading NOAA climatologist, says that recent warming affects nighttime temperatures more than daytime temperatures, and that the data show less frequent but heavier rainfall—perhaps a trend in storminess. Even assuming a doubling of carbon dioxide to stoke the greenhouse effect, future temperature trends are even less certain. One of meteorology's gravest concerns is the role of hurricanes in global warming. Hurricanes exploit temperature differences between the bottom and top of the troposphere, pumping heat toward midlatitudes. The hurricanes are likely to be even stronger if the temperature difference between the ocean surface and the top of the troposphere increases. In the 1980s, Kerry Emanuel of MIT showed that theoretically global warming could significantly increase hurricane intensities. In 1998, a team of hurricane modelers at NOAA's Geophysical Fluid Dynamics Labs in Princeton, New Jersey, confirmed this once-controversial claim with computer modeling. They simulated sea temperature increases with their global climate model, then plugged these results into the GFDL hurricane development model, which is currently the top model used by the National Hurricane Center. The GFDL scientists say the strongest hurricanes could spin 5 to 12 percent faster with a 4 degree rise in sea surface temperatures. That's a devastating increase, unless, paradoxically, the catastrophic rise in sea level somehow reconfigures the coastline to blunt the force of storm surges.

While some meteorologists are already stretching the limits of the science—peering far into the future and thinking on a global scale—others are perplexed by storms that refuse to fit into accepted categories. In this century meteorologists finally codified the differences between tropical and extratropical cyclones. Tropical cyclones, driven by moist rising air, transfer solar energy directly to the upper troposphere for delivery to midlatitudes. As heat engines, they sort out vertical temperature contrasts but thrive in warm uniformity. Extratropical cyclones, on the other hand, sort out the horizontal temperature contrasts between polar and tropical air. They are asymmetric and exploit the clash of air from opposite sides of fronts.

Satellite images consistently reveal exceptions, however. Plenty of storms fit neither the spiral of the hurricane nor the comma-shaped cloud canopy of the extratropical storm. Some of the oddest storms appear north of the Arctic Circle, near the polar ice cap. These winter Arctic Ocean cyclones have spiral bands and a clear eye just like a hurricane, but clearly don't require the extravagant warmth of a hurricane. These polar lows form when −40-degree air flows off the ice cap over the (relatively) warm ocean. The ocean warms the cold air flowing off the ice, creating a volatile instability. Whereas in a hurricane evaporation of the sea surface is critical, the polar low is so cold that little ocean water evaporates. Nonetheless, the temperature difference between the water and air is so great that the inflow in a polar low packs a punch of potential energy similar to that of a tropical hurricane. Convection results, forming a low-pressure center and gathering winds from all directions. The rotation of the Earth bends winds sharply near the poles (much more strongly than in the tropics) to make a cloudy spiral and an eye. With a low troposphere ceiling in the Arctic, convection reaches only a few miles high—less than half the height of a hurricane.

The strongest of these polar lows are not impostors—they're dangerous storms. Scientists have begun calling them Arctic hurricanes. The tightly wound circulation is very compact. At 100 miles across it is much smaller than most hurricanes, but the winds can exceed 74 m.p.h., just like a real hurricane. Also, in the Arctic, the wind is wet and cold enough that ice accumulates on ships, making many vessels perilously top-heavy. Arctic hurricanes are particularly dangerous because they mature in a day or less—several days or even a week faster than a tropical cyclone. They also speed toward landfall at close to 30 m.p.h., much faster than the average hurricane.

Satellite images also have revealed the ominous spiral of hurricanelike storms in the Mediterranean. One lashed a German ship with 84-m.p.h. winds. The sea surface temperatures were nearly 20 degrees cooler than necessary to make a traditional hurricane, but thunderstorms nonetheless banded together into a spiral around the eye.

Off the coasts of the United States and elsewhere, forecasters keep a close watch on extratropical storms that develop eyes. These hybrid storms are among the most damaging. The infamous Halloween nor'easter of 1991 pitched 100-foot waves in the open Atlantic and caused hundreds of millions of dollars in damage along the East Coast. As a hybrid, it developed an eye and distinct bands when it dipped its toes into the warm Gulf Stream for a day. The National Hurricane Center sent its airplane into the storm, where scientists found telltale signs of tropical development, including 100-m.p.h. winds (at 3,000 feet) around a core of air 7 degrees warmer than its surroundings. The storm was purely extratropical in structure—boasting fronts and a cold core—and the strongest winds spread out along the wide-ranging cold front. During its brief metamorphosis into a tropical-style hybrid, however, the Halloween storm masked its true identity by pulling its maximum winds to within 30 miles of the center, just like a hurricane would in an eyewall.

In 1994, National Hurricane Center specialists watched again as a confused storm started up along a frontal boundary near Florida. It was four days before Christmas—several weeks after the end of the hurricane season—yet the storm wound its winds and clouds into a diffuse spiral. Though the winter holidays are a season when storms die if their frontal contrasts weaken, this storm continued strengthening over the Gulf Stream despite losing its fronts all together. Satellite images suggested that the storm was a tropical storm, not a nor'easter, as it battered Cape Cod on Christmas Eve with 100-m.p.h. winds.

The warm core within a hybrid storm defies identification. Even if an eye forms, it is not cleared out by sinking air from eyewall thunderstorms the way a hurricane eye is. NOAA scientist Mel Shapiro suggests that air from the warm sector of the storm slides past the old low center and ends up west of the low. The warmth around the old low means that the warm and cold fronts may not meet; the cold front falls short of the warm front. The fronts make a fractured T-bone structure (the cold front is the broken stem), with a "secluded" mass of warm air at the left tip of the T. Around the secluded warm air, thunderstorms may crop up, embedded within the original comma-shaped cloud of the extratropical storm. Such storms can intensify quickly, in part because they have both extratropical frontal contrasts and a tropical-style concentration of latent heat release.

Meteorologists would never have noticed confused storms like these without satellites. The warm lows embedded in hybrid storms are nearly impossible to locate with the usual balloon soundings and ship's reports. Meteorologists also probably wouldn't have been able to analyze hybrids confidently if they hadn't already moved far beyond the Bergen model of extratropical storms, at least in research settings. For instance, the late stage of cyclone maturity—occlusion—endured a great deal of scrutiny from the beginning. The cold front, usually described as ideally a wedge of air spreading farthest at the bottom, nonetheless often catches the warm front aloft first, rather than at the ground level.

In the 1950s, Canadian meteorologists, dissatisfied with the Bergen model, came up with a "three-front" storm analysis technique. The Canadians said that after occlusion, when the cold front catches the warm front and erases the contrasts in the storm, the leftover surface front basically becomes benign. The real weather of such a mature storm instead forms along a midlevel low-pressure trough parallel to and ahead of the surface occluded front. In this "trough aloft," the warm air sector nestles atop the old cold front. Air rising through the trough aloft creates some of the most significant precipitation of the storm. A University of Wisconsin meteorologist recently revived aspects of this analysis technique to show how an otherwise moderate January 1995 cyclone in the Midwest developed an intense band of snow 600 miles long under a trough aloft.

Similarly, since World War II, East Coast forecasters have suspected that snowstorms along the Atlantic Coast—many of them classic nor'easters—don't strictly follow the Bergen model. Half the storms analyzed in Paul Kocin and Louis Uccellini's recent landmark book *Snowstorms Along the Northeastern Coast of the U.S.*, underwent an abrupt redevelopment that enabled them to leapfrog over the Appalachians. As a storm approached the mountains, a secondary low-pressure center developed along the warm front east of the mountains. The secondary low was intensified by the cool air along the eastern slope of the mountains meeting the warmer ocean air and carried on for the earlier low center. The storm overall jumped eastward to a new parallel track, just like the Blizzard of '96 over the Carolinas. Most of the other storms in the historical study by the two National Weather Service meteorologists also jumped ahead to a new low near the coast, but stayed along a consistent track. In either case, the new storm centers developed sharp fronts when the warm air sweeping onto the coast temporarily "dammed" the cold air against the mountains.

Pointing out that forecasters frequently have no choice but to force unusual storms into Bergen-style analysis—like square pegs into round holes—Clifford Mass of the University of Washington recommended changes be made in the model. "Unfortunately," he admitted in a 1991 paper, "a comprehensive conceptual model integrating the insights gained over the past half-century does not exist." Given the variety of situations and features observed over that time, it is entirely possible that the days of one all-encompassing extratropical storm model are over. Mass listed many lessons learned since 1918. Whereas the Bergen model suggests that a semipermanent polar front undulates to move preexisting cold and warm fronts around growing low-pressure centers, in fact the fronts and the cyclone center develop more-or-less together. Also, the warm front appears to wither away altogether during the waning hours of an extratropical system. Similarly, Tor Bergeron long ago helped define the end of the storm's life cycle with occlusion, but very few meteorologists have recorded a cold front catching the warm front. Sometimes occlusion seems to be instantaneous: the warm air appears aloft without explanation.

Meteorologists have extended the life of the Bergen model by adding on bit by bit as if they were building a rambling vacation villa. To express the three dimensions of motion in a storm, analysts now often talk about conveyor belts of air instead of fronts. They've turned the Bergen model into a tangle of colorful viewgraph spaghetti. A stream of air called the "warm conveyor belt" rises slowly northward (in the northern hemisphere) along the cold front, then after passing the warm front soars northeast to midlevels of the storm. A cold conveyor belt crosses underneath the warm belt north of the warm front. The cold conveyor belt evaporates some of the copious precipitation falling from the warm belt above, but leaps upward over the low center, dumping snow or rain of its own. Most of the cold conveyor belt doubles back like a limbo dancer, its head tilted eastward with the jet stream. Some of the cold belt, however, hooks around the low and spreads behind the cold front. Here it mingles with a dry airstream swooping down from high altitudes.

In 1998, NOAA researchers took advantage of El Niño by stationing hurricane hunting aircraft in Monterey, California, for an intensive study of coastal storms. Mobile Doppler radars from the University of Oklahoma were also there getting fine scale readings of the landfalling winds and rains. The scrutiny proved very useful, in one case allowing forecasters to make a pinpoint flash flood warning six hours earlier than otherwise possible. The researchers also found new structures in the storms that explained their vehemence. As some storms reached shore, their cold air could not scale the coastal mountains. Instead they spread north and south, creating a frontlike ridge of air along shore. Subsequent storms then slid over this dense air, dropping copious rains ahead of schedule. In addition, the wettest air rode in on low-level jets. One raced onshore at 110 m.p.h. at about 2,000 feet in altitude. The confirmation of these jets, one of the goals of the project, shows why many of the storms localized their floods and damaging winds. The Bergen model of storms, naturally, includes no such features.

Descriptive, yet mathematical, analyses of the movement of air along conveyor belts have crashed the study of storms headlong into the most intriguing question that meteorologists must face in the next century. The situation is very similar to the last turn of the century. No one could have predicted that meteorologists, who began the 19th century without knowing which way the wind blew in most storms, would end it with thermodynamics-based analyses of storms. That remarkable progress led Cleveland Abbe and Vilhelm Bjerknes and others to think 20th century meteorologists might finally perfect weather prediction by doing it mathematically. According to Tor Bergeron, in the early 1900s Vilhelm Bjerknes became obsessed with weather prediction by rational means. Not by "cataloguing and memorizing, of isobaric patterns and weather types … but as a mathematically well defined physical problem." Bjerknes hoped for adding machines that could circumvent the difficult calculus involved in the basic equations of heat and energy and motion involved in weather.

Two brilliant men did not wait for such a machine. One was Felix Exner, a professor at the University of Vienna. In a 1908 paper he described his calculations of air motion, yielding surface temperature and pressure changes. But his method was exceedingly crude, especially since he worked at a time when few observations of upper winds were available. He capped his model at 16,000 feet and erroneously assumed that upper winds moved identically to surface winds. Exner nonetheless may have inspired the English mathematician Lewis Fry Richardson to give numerical forecasting a try.

Like many of today's meteorologists, Richardson first showed interest in meteorology as a mathematician. While working for the peat industry on draining bogs, Richardson had developed step-by-step methods for approximating the kinds of smooth, continuous flow problems normally reserved for calculus. By 1907 the Cambridge-educated mathematician decided that weather prediction was as well suited to his new mathematical techniques as peat bogs. Richardson started the project the year before World War I broke out, rewriting the basic equations behind weather as a simpler arithmetic. A Quaker, he refused to fight in the war but instead volunteered as an ambulance driver in France. Under the stressful conditions working near the front, Richardson solved the equations necessary to move the atmosphere for six hours across a grid of data points representing Europe. It took him six weeks of exhausting hand calculations. The results were ludicrous: where nature had made a very small drop in pressure (the test case of May 20, 1910), Richardson had made a superhurricane with winds of more than 200 m.p.h. on paper. Undaunted, he published the work in 1922 in one of the great books of meteorology, *Weather Prediction by Numerical Process.* He dreamed of a day when forecasters would build a round arena with many tiers of offices, in which thousands of people would gather, each making calculations representing one point on the Earth's surface, to create continuous perfect weather forecasts.

Later, Richardson's colleagues realized that widespread data points had ruined his forecast. Special treatments for this mathematical snag had yet to be discovered. But the basic methods Richardson used had brought meteorology to the brink of objective weather forecasting. The global weather-calculating office he envisioned shrank to the silicon-chip-packed plastic boxes of today's fastest supercomputers. The computer was developed for the military and originally dedicated for projects like gunshot trajectories. But John von Neumann, the brainy and charismatic mathematician who invented the first programmable computer, had other ideas as well. He wanted to see science get a piece of the computer pie and demonstrate the capabilities of his new machine.

One can imagine that a meeting with Carl Gustav Rossby in 1942 helped turn von Neumann's thoughts toward the mathematical problems of meteorology. It must have been impressive, the sizzling synergy between Rossby, the von Neumann of modern meteorology, and von Neumann, the Rossby of modern applied mathematics. A more likely theory is that RCA engineer Vladimir Zworykin suggested the idea. He engineered a meeting between von Neumann and Francis Reichelderfer, head of the

Weather Bureau, in late 1945, and soon von Neumann was eager to have his computers tackle what he believed was "the most complex, interactive, and highly nonlinear problem that had ever been conceived of—one that would challenge the capabilities of the fastest computing devices for many years." He was right on all counts.

Within a couple of months, with Rossby's encouragement, von Neumann established a meteorology group at the Institute for Advanced Study at Princeton, where he was working on the first stored-program electronic computer. Progress on the forecasting problem was slow, however, until Jule Charney, a freshly minted Ph.D. in meteorology from UCLA, joined the project. Building on Rossby's work with high-altitude waves and jet streams, Charney had taken stock of Richardson's equations and had seen that much of the atmospheric wave action they described was largely superfluous to modeling the weather. So he filtered out effects like sound and gravity waves. He decided the best way to tackle the problem for the time being was to approximate an idealized atmosphere similar to the low friction flow at midaltitudes, where the rotation of the Earth and pressure gradients in the air cancel one another out. In this brilliant approximation, he reduced the several equations down to one. Other innovations necessary to make a forecast with the limited computers followed.

Since von Neumann's first computer wasn't yet ready in 1950, Charney and his colleagues went to the Army's Aberdeen Proving Ground, where the earlier ENIAC computer was located. It took more than a month's work to prepare the punch cards and circuits for six test forecasts, but the results were encouraging. Further tests at Princeton seemed to prove the promise of the computer age—von Neumann's new computer took only five minutes to predict a day's worth of atmospheric motion. By 1953, Rossby had moved back to Sweden, where he established a computer forecasting group that made the first 24-hour forecasts. Handling the punchcards, running the computer, and analyzing the results took six hours, but a new age of meteorology was at hand. Reichelderfer soon started a similar forecasting group for the Weather Bureau in Washington, and by 1955 it was making the first daily forecasts by computer.

Meteorology has remained at the vanguard of the computer era. Each generation of supercomputers is eagerly snapped up by weather forecast services and research institutions. Each new computer means capacity for more realistic models. Eventually, computers were fast enough to bring back some of the detail trimmed by Charney. Nonetheless, it wasn't until the 1970s that computers finally began to predict weather as well as the best human forecasters. And "weather" to a computer then really meant only the upper-air patterns steering the rain, heat, and winds below. In the last decade, researchers have tested sophisticated new models with resolution fine enough to include terrain features that are essential to making and breaking storms. Individual thunderstorms still slip through the grid, except in high-powered regional models, such as the research mesoscale forecast model at the University of Oklahoma that calculates the weather at the scale of individual clouds.

The computer age created a meteorology bristling with confidence. But for all the many conceptual advances made by using computers— modeling supercells, hurricanes, rainclouds, and jet streams—the most significant discovery of the computer age may very well have been the cautionary results produced by one MIT meteorologist with a Royal McBee computer in the late 1950s. The computer often sputtered, requiring Edward Lorenz to rerun simulations. One day he rekeyed the conditions during a simulation of the atmosphere and started up the computer. While Lorenz went out for a coffee break, the computer began to spit out bizarre answers. This happened often, and when he returned, Lorenz thought the machine was breaking down. But soon he realized that the problem this time was in the initial conditions. In retyping them, he had entered rounded numbers from the printout, not the actual numbers that had been stored in the computer. This small change in initial conditions had produced a wildly different answer. It was chaos.

The weather is chaotic too. Storms come one day and not the next simply because of small changes in wind directions or pressures. The tiniest differences between barometer readings and reality, between inevitable computational shortcuts and nature, eventually make huge differences between simulation results and the real world.

Chaos troubles forecasters with the National Weather Service. The supercomputer gurus at the National Centers for Environmental Prediction recently began running models more than once. Just as Lorenz did, they change the initial conditions a little bit here and there, trying to see how chaotic the atmosphere is behaving on a given day. Sometimes the model results diverge wildly; sometimes they are relatively consistent. In 1998, NOAA sent storm experts flying out over brewing Pacific disturbances near Alaska to get vastly improved initial observations of the systems heading toward the United States. They found that computer models consistently underrepresented the winds. In one case the scientists recorded 240-m.p.h. winds in the jet stream, faster than any hurricane. One February weekend looked stormy for Louisiana, given the initial model outputs, but the scientists dropped instrument packets from the high-flying Gulfstream jet and recorded extra intensity in the Pacific winds. The effort paid off when forecasters reran the supercomputer models with the new data. The jet stream looked like it would swing more violently south and in a few days head over Florida. That's exactly what happened. Huge V-shaped anvils cropped up above powerful thunderstorms over central Florida under a streak of 145-m.p.h. jet-level winds. Despite a long tornado watch, 38 people died near Orlando in a nighttime tornado outbreak. One of the twisters may have been only the fifth F4 ever in the state.

Given that the slightest wiggles in flow over the Pacific can grow into major U.S. disasters, the sparse data available there is an acute problem. The University of Washington is planning to fly 10 robotic planes with 1,000-mile range out over the ocean on preprogrammed flights to fill the gaps in the oceanic observations. Automated planes of this type may someday fill the gaps in many areas of meteorology, from ozone observations to storm chasing at high altitude. They may also contribute in hurricane forecasting, where precise intensity, position, and environmental data are crucial to better intensity and track model results. Not surprisingly, hurricane development and intensification is one of the most difficult problems in the field in part because it is chaotic. The sensitive dependence of the hurricane's life cycle on the small asymmetries cropping up in its structure appeared before 1970 in mathematical modeling by Katsuyuki Ooyama. More recently, Greg Holland, a researcher for the Australian Bureau of Meteorology, has investigated this chaos in the interaction between the small vortices of the hurricane and the larger, monsoonal vortices of its environment. Similarly, the wild variations of the El Niño cycle are a sign of its chaotic origins, an interaction of numerous cycles sensitively dependent on initial conditions.

Better observations won't cure forecasts of the ills of chaos. Lorenz showed that no degree of accuracy is high enough when nature is chaotic. The laws of weather make it chaotic. This was a stunning revelation to many scientists. Not because they didn't suspect that it was true, but because all of their effort had gone into eliminating the chaos of weather. Ever since the mid-19th century, when thermodynamics entered the study of storms, meteorologists had hoped for a day when predictions of weather would look as good as the astronomers' prognostications of tides and eclipses.

In their greatest hour, meteorologists ironically have discovered their limits. The rush of new discoveries about sprites, El Niño, and other phenomena have forced meteorologists to look 100 miles higher and thousands of miles wider for a proper perspective on the clouds overhead. Sudden storm intensification, oddball tornadoes, and hybrid cyclones have blurred the lines meteorologists have been drawing. They present a hitherto unsuspected confusion in weather.

No one who knows weather is surprised by this, for confusion is synonymous with weather. Benjamin Franklin saw through much of the confusion in the dark, roiling clouds above him. He fitted humanity with its first conceptual glasses for focusing on the order in the chaos of the atmosphere. Predictably, they were bifocals. Through one half we saw the weather as physics incarnate, and meteorologists pursued long-range goals like objective storm forecasts. Through the other half, however, we read nature's threatening but delightfully endless creativity with water, ice, and wind. It spelled our mortality, such as in the Blizzard of '96, as well as our relentless fascination with forces beyond our control.

After two extraordinary centuries of scientific development, meteorologists at last grapple with the weather in the sky itself or play by its own rules in the safety of parallel processors. They endlessly grind and polish Franklin's conceptual lens on the weather, yet they cannot turn our gaze from the beauty and power of wild weather. We are still blessed with an intuitive compulsion for experiencing the coexistence of the eternal and the ephemeral in storms. Knowing what they know, seeing what they see, scientists help us retain the sense of wonder and awe of truth that led Benjamin Franklin to launch a kite and Cotton Mather to pray for forgiveness. Meteorologists have transformed our world into a vision of splendor that we can grasp but never own, an insight into complexity that we can appreciate but never fully comprehend. Face to face with the weather, our capacity to imagine expands but our ability to understand is never sufficient. CBS newsman Edward R. Murrow observed this when he flew with the hurricane hunters to witness firsthand the magnificence of storms. "If an adequate definition of humility is ever written," he said, "it's likely to be done in the eye of the hurricane."

Notes

CHAPTER ONE: A STORMY RELATIONSHIP

1   In a few short days ... "Blizzard-Staggered Government Shut Again," *USA Today* Online (14 January 1996).

2   This was the scene ... Douglas LeComte, "A Wet and Stormy Year," *Weatherwise* (February/March 1997), 16.

2   After nearly a day and a half ... Desda Moss, "Effects of the Blizzard," *USA Today* Online (9 January 1996).

2   More than 40 inches ... Paul J. Kocin, William Gartner, and Daniel H. Graf, "Snow: A Record Season," *Weatherwise* (February/March 1997), 32–33.

3   At least 33 people drown ... LeComte, "A Wet and Stormy Year," 16.

3   Weather experts have a bone to pick ... Paul Knight, "Outtakes," *Weatherwise* (February/March 1997), 37.

4   This storm smashed ... Brad Rippey, "Weatherwatch," *Weatherwise* (April/May 1996), 42.

5   Kublai Khan found out the hard way ... Richard Williams, "The Divine Wind," *Weatherwise* (October/November 1991), 11–14.

5   This time history is unambiguous ... H. Arakawa, "The Weather and Great Historical Events in Japan," *Weather* (May 1960), 155.

5   The Spaniards intended to dispose ... H. H. Lamb, "The Weather of 1588 and the Spanish Armada," *Weather* (November 1988), 386–394.

7   Today there are nearly 100 ... David D. Houghton, "Meteorology Education in the United States after 1945," in James Rodger Fleming, ed., *Historical Essays on Meteorology, 1919–1995* (Boston: American Meteorological Society, 1996), 547.

8   In 1938 a hurricane off the Florida ... , David M. Ludlum, "The Great Hurricane of 1938," *Weatherwise* (August/September 1988), 215–216.

9   The next day, people on the Long Island shore ... Joe McCarthy, *Hurricane* (1969).

9   In 1938, had forecasters viewed a map ... A. James Wagner, "The Impossible Hurricane," *Weatherwise* October/November 1988, 279–283.

10   If you make a quick tally ... Patrick Hughes and Stanley David Gedzelman, "Meteorology and America, 1920–1995," *Weatherwise* (June/July 1995), 26–70.

10   D-Day itself hinged on its weather ... R.A.S. Ratcliffe, "Weather Forecasting for D-Day, June 1944," *Weather* (June 1994), 198–202.

11   The fleet, commanded by Admiral ... Hans C. Adamson and George F. Kosco, *Halsey's Typhoon* (New York: Crown Publishers, Inc., 1967), 1–4, 25–121.

14 Writing in the *New York Tribune* ... William Ferrel, *A Popular Treatise on the Winds* (New York: John Wiley and Sons, 1893), 425–426.

15 Not by accident did Torricelli write ... Karl Schneider-Carius, *Weather Science, Weather Research: History of their Problems and Findings From Documents During Three Thousand Years*, trans. Indian National Scientific Documentation Centre, New Delhi, (NOAA, 1975), 72.

16 Clouds, he wrote, were not ... *Ibid.*, 150.

16 He wrote: "The grouping ... *Ibid.*, 151.

17 Howard, born in 1772 ... Brian J. R. Blench, "Luke Howard and His Contribution to Meteorology," *Weather* (March 1963), 83–92.

17 He proposed three basic categories ... Blench, "Luke Howard and His Contribution to Meteorology," 88–89.

17 One German translation ... John A Day and Frank Ludlam, "Luke Howard and His Clouds: A Contribution to the Early History of Cloud Physics," *Weather* (November 1972), 452.

18 Landscape artists like John Constable ... L.C.W. Bonacina, "Turner as a Weather Painter," *Weather* (July 1951), 200–203.

19 The local NBC affiliate ... Marc Fisher, "The Storm Trooper," *Washington Post* (8 January 1996), D1.

19 The storm, we could all see ... Lee Grenci, "A Smash Hit Broadway Show," *Weatherwise* (April/May 1996), 48.

CHAPTER TWO: A LIGHT IN THE DARK

21 "Good God! What horror ... Patrick Hughes, "Hurricanes Haunt Our History," *Weatherwise* (June/July 1987), 138.

21 Wrote Increase Mather ... Increase Mather, *Remarkable Providences Illustrative of the Earlier Days of the American Colonisation*, (Boston, 1684, reprinted in London: Reeves and Turner, 1890), 88.

21 In an exhaustive study ... *Ibid.*, 51.

21 "[A]lthough terrible lightnings with thunders" ... *Ibid.*, 51.

21 Sometimes people seem to see nothing ... Mary Reed, "And Not One Vicious Liver in the Lot ... ," *Weatherwise* (June/July 1988), 172–174.

22 It was also possible to see storms ... Randy Cerveny, "Power of the Gods," *Weatherwise* (April/May 1994), 22–23.

23 "Perhaps I should on this occasion" ... Nathan G. Goodman, ed., *The Ingenious Dr. Franklin: Selected Scientific Letters of Benjamin Franklin*, (Philadelphia: University of Pennsylvania Press, 1956), 1.

23 But the lightning rod also represents ... I. Bernard Cohen, *Benjamin Franklin's Science* (Cambridge: Harvard University Press, 1990), 6.

23 Franklin didn't fully understand ... *Ibid.*, 14

23 Franklin was in good company ... *Ibid.*, 18.

23 He tested various clothing ... Goodman, *The Ingenious Dr. Franklin*, 11.

25 For these discoveries ... Cohen, *Benjamin Franklin's Science*, 4.

25 "[The demonstrations] were imperfectly ... " Benjamin Franklin, *The Auto-biography of Benjamin Franklin*, in Raymond Phineas Stearns, *Science in the British Colonies of America* (Urbana: University of Illinois Press, 1970), 620.
25 His neighbors were so delighted ... , *Ibid.*, 621–622.
26 The ancient Greeks recognized ... Hal Hellman, *Light and Electricity in the Atmosphere* (New York: Holiday House, 1967), 17.
26 In 1708, William Wall noticed ... *Ibid.*, 40.
26 Nine years later, Sir Isaac Newton ... *Ibid.*, 40.
27 "Are flashes and sparks ... " Schneider-Carius, *Weather Science, Weather Research*, 100.
27 Ideas about what the atmosphere ... S.K. Heninger, Jr., *A Handbook of Renaissance Meteorology with Particular Reference to Elizabethan and Jacobean Literature* (Durham: Duke University Press, 1960), 3.
27 Rather than investigate weather ... W.E. Knowles Middleton, *A History of the Theories of Rain and Other Forms of Precipitation* (New York: Franklin Watts, Inc., 1965), 10.
28 Others who took up ... Heninger, *A Handbook of Renaissance Meteorology*, 12.
29 Editions of Greek scholarship ... *Ibid.*, 10.
29 First, Aristotle adopts Eudoxus's ... H. Howard Frisinger, *The History of Meteorology: to 1800* (New York: Science History Publications, 1977), 15.
29 Even Johannes Kepler ... Schneider-Carius, *Weather Science, Weather Research*, 62–63.
29 Christopher Columbus (according to his nephew Ferdinand) ... Clive Hart, *The Prehistory of Flight* (Berkeley: University of California Press, 1985), 16.
30 "A cloud is kindly hollow" ... Heninger, *A Handbook of Renaissance Meteorology*, 49.
30 In the 1400s, William Caxton ... *Ibid.*, 49.
30 Leonardo Da Vinci, whose paintings ... Stanley David Gedzelman, "The Sky in Art," *Weatherwise* (December 1991/January 1992), 8; and Schneider-Carius, *Weather Science, Weather Research*, 58.
30 Bacon wrote ... Middleton, *A History of the Theories of Rain*, 18.
30 William Fulke used hot-and-cold contrasts ... Heninger, *A Handbook of Renaissance Meteorology*, 21
31 But an even less worthy ... *Ibid.*, 51–55.
32 In the regional rank ... *Ibid.*, 108.
32 As for the strong winds ... *Ibid.*, 118–119.
32 Bartholomeus saw that north ... *Ibid.*, 114.
32 The exhalations burst ... *Ibid.*, 73.
32 Like other Renaissance thinkers, *Ibid.*, 75–76.
33 Similarly, Scandinavians ... Cerveny, "Power of the Gods," 21.
34 And Thomas Hill ... , Heninger, *A Handbook of Renaissance Meteorology*, 80–82.
34 Various versions of the story ... Andrew Dickson White, *A History of the Warfare of Science with Theology in Christendom* (New York: Macmillan and Co, 1896, reprinted in New York: Dover, 1960), 332.
34 A book published in Zurich ... *Ibid.*, 335.
35 Commented the Bishop of Voltoraria ... , *Ibid.*, 333.
35 Baptismal instructions published ... , *Ibid.*, 346.

35 Many church bells ... , *Ibid.*, 345.

35 Another boasted ... , *Ibid.*, 345.

35 Thomas Aquinas wrote ... *Ibid.*, 337.

35 Skeptics warily pointed out ... *Ibid.*, 339.

36 In Newton's time, the rector ... *Ibid.*, 350.

36 On the one hand you have Increase ... *Ibid.*, 335.

36 On the other hand you have Increase ... , Stearns, *Science in the British Colonies of America*, 154–155.

36 Cotton Mather was a rare ... Silvio A. Bedini, *Thinkers and Tinkers: Early American Men of Science* (New York: Charles Scribner's Sons, 1975), 77–82

36 Increase commented wryly ... , *Ibid.*, 82.

37 "The electrical fluid is attracted" ... Stearns, *Science in the British Colonies of America*, 623–624.

37 So Franklin proposed ... Cohen, *Benjamin Franklin's Science*, 4.

37 Franklin and his son, James ... *Ibid.*, 67–69.

38 On May 10, 1752 ... Hellman, *Light and Electricity*, 42–44.

38 Within the year ... , Schneider-Carius, *Weather Science, Weather Research*, 103.

39 In America, a Baptist minister ... Stearns, *Science in the British Colonies of America*, 509–510.

39 In Charleston, South Carolina, a newspaper ... Cohen, *Benjamin Franklin's Science*, 145.

39 San Marco in Venice had been damaged ... Cerveny, "Power of the Gods," 23.

39 Prince misunderstood Franklin's discovery ... , Cohen, *Benjamin Franklin's Science*, 145.

40 The good Knight was by virtue ... Stearns, *Science in the British Colonies of America*, 600.

CHAPTER THREE: STORMS ON THE MOVE

41 Rife with the usual follies of battle and diplomacy ... Peter Gibbs, *Crimean Blunder: The Story of War with Russia a Hundred Years Ago* (New York: Holt Rinehart and Winston, 1960), passim.

42 Le Verrier at once wrote ... Helmut Landsberg, "Storm of Balaklava and the Daily Weather Forecast," *Scientific Monthly* (December 1954), 347–352.

43 As the science historian James Fleming ... James Rodger Fleming, *Meteorology in America, 1800–1870* (Baltimore: Johns Hopkins University Press, 1990), xx–xxii.

43 FitzRoy's new job, for instance ... Landsberg, "Storm of Balaklava," 350–352.

44 "This puzzled me," Franklin later explained ... Goodman, *The Ingenious Dr. Franklin*, 185–187.

45 The first observation of the whirlwind ... Mark Cherrington, "The Pirate Scientist," *Weatherwise* (August/September 1989), 205–207.

45 Franklin said that tropical heat thins ... Goodman, *The Ingenious Dr. Franklin*, 187.

46 Redfield's father, a sailor ... Denison Olmsted, "Biographical Memoir of William C. Redfield," *American Journal of Science* (1857), 355–373.

46 Next it wreaked havoc ... David M. Ludlum, *Early American Hurricanes, 1492–1870* (Boston: American Meteorological Society, 1963), 83.

47 Redfield noted that experts ... William C. Redfield, "Remarks on the Prevailing Storms of the Atlantic Coast," *American Journal of Science* (1831), 17–51.

48 No less a figure than ... Olmsted, "Biographical Memoir," 364.

49 One brig, the *Charles Heddles* ... Ivan Ray Tannehill, *The Hurricane Hunters* (New York: Dodd, Mead & Co., 1955), 28–30.

50 "In no department" ... Olmsted, "Biographical Memoir," 364–365.

51 Not long after, his contemporary ... Schneider-Carius, *Weather Science, Weather Research*, 79–82.

51 Heinrich Brandes of Prussia ... Landsberg, "Storm of Balaklava," 348–350.

52 Then, for a severe storm ... Gisela Kutzbach, *The Thermal Theory of Cyclones: A History of Meteorological Thought in the Nineteenth Century* (Boston: American Meteorological Society, 1979), 31.

52 The telegraph operators were not trained ... Patrick Hughes, "The Great Leap Forward," *Weatherwise* (October/November 1994), 22.

52 In many ways Henry's early life ... Frank Millikan, "A Self-Made Scientist," *Weatherwise* (October/November 1997), 16–17.

53 "I did not then consider it" ... Thomas Coulson, *Joseph Henry, His Life and Work* (Princeton: Princeton University Press, 1950), 64.

53 Henry, on the other hand, won ... Patrick Hughes, "Keepers of the Flame," *Weatherwise* (October/November 1997), 20.

53 Each observer received ... Frank Millikan, "Joseph Henry's Grand Meteorological Crusade," *Weatherwise* (October/November 1997), 14–18.

54 If the information didn't serve ... *Ibid.*, 16.

54 In 1868 and 1869 ... Hughes, "The Great Leap Forward," 26.

54 Now the Army's Signal Service ... Charles C. Bates and John F. Fuller, *America's Weather Warriors, 1814–1985* (College Station: Texas A&M University Press, 1986), 10.

54 On November 8 ... Truman Abbe, *Professor Abbe and the Isobars* (New York: Vantage Press, 1955), 123–124.

55 "The science of meteorology" ... *Ibid.*, 101.

55 "Clouds and warm weather" ... *Ibid.*, 3–4.

56 His success as a forecaster ... *Ibid.*, 5.

56 The climatologist Helmut Landsberg ... Landsberg, "The Storm of Balaclava," 352.

57 "It should always be remembered" ... Patrick Hughes, "FitzRoy, Forecaster," *Weatherwise* (August 1988), 200–205.

CHAPTER FOUR: AN ENLIGHTENED CONFUSION

59 "A very dense and low cloud" ... Lewis C. Beck, "Note on the New Brunswick Tornado or Water Spout of 1835," *American Journal of Science* (1840), 115.

59 "In a few minutes" ... *Ibid.*, 116.

60 "To establish the bare fact" ... Nathan Reingold, ed. *Science in Nineteenth-Century America: A Documentary History* (New York: Hill and Wang, 1964), 104.

60 They saw the same piles of bricks ... Fleming, *Meteorology 1800–1870*, 32–33.

60 Despite eyewitness accounts of rotation ... William C. Redfield, On the Courses of Hurricanes; with Notices of the Tyfoons of the China Sea, and Other Storms," *American Journal of Science* (1839), 201–227.

61 "Indeed, meteorology has ever been" ... David M. Ludlum, "The Espy-Redfield Dispute," *Weatherwise*, (December 1969), 224.

62 Meteorology emerged with central concepts ... Kutzbach, *The Thermal Theory of Cyclones*, 2–3.

62 He found the experiments of John Dalton ... *Ibid.*, 20–22.

62 Otto von Guericke vividly demonstrated ... Middleton, *A History of the Theories of Rain*, 44–45.

62 In 1648 the French mathematician ... Schneider-Carius, *Weather Science, Weather Research*, 75–79.

62 In 1660, English philosophers ... Robert D. Purrington, *Physics in the Nineteenth Century* (New Brunswick and London: Rutgers University Press, 1997), 76–77.

63 Franklin later mused ... Goodman, *The Ingenius Dr. Franklin*, 203–208.

63 Hadley calculated that air ... Schneider-Carius, *Weather Science, Weather Research*, 96.

64 He matched convection with a crucial ... Richard Williams, "The Mystery of Disappearing Heat," *Weatherwise* (August/September 1996), 28–29

65 The result was an instantaneous transition ... Kutzbach, *The Thermal Theory of Cyclones*, 24.

65 Adiabatic expansion and contraction ... *Ibid.*, 22.

66 Some of Redfield's ideas ... Ludlum, "The Espy-Redfield Dispute," 224.

66 In one of his most damaging claims ... Fleming, *Meteorology in America, 1800–1870*, 36–37.

67 "The grand error into which" ... Ludlum, "The Espy-Redfield Dispute," 225.

67 The idea that a central updraft ... *Ibid.*, 225.

67 In 1844, aware that Hare ... Fleming, *Meteorology in America 1800–1870*, 50.

67 Hare "immediately rose" ... *Ibid.*, 96.

67 And yet, in another meeting ... *Ibid.*, 97.

68 Redfield declined, saying he ... Ludlum, "The Espy-Redfield Dispute, 229.

68 One newspaper reported that ... *Ibid.*, 229.

68 Another swooned ... *Ibid.*, 229.

68 He suggested torching a line of forestland ... Fleming, *Meteorology in America 1800–1870*, 99.

69 He wisely believed that ... *Ibid.*, 49.

69 He retired and went to Ohio ... Ludlum, "The Espy-Redfield Dispute," 245.

69 One called Espy a "monomaniac" ... Clark C. Spence, *The Rainmakers: American "Pluviculture" to World War II*, (Lincoln and London: University of Nebraska Press, 1980), 16.

69 His friend, Alexander Bache ... Kutzbach, *The Thermal Theory of Cyclones*, 43.

69 Redfield and Espy disagreed ... Gregory De Young, "The Storm Controversy (1830–1860) and its Impact on American Science," *Eos* (17 September 1985), 657–660.

70 Isobars—lines of equal pressure ... Kutzbach, *The Thermal Theory of Cyclones*, 32–33.

70 Loomis believed synoptic maps ... *Ibid.*, 71.

71 For Loomis, the clash ... *Ibid.*, 29–30.

72 Dove tracked the wind shifts ... Schneider-Carius, *Weather Science, Weather Research*, 217–219.

72 "The dependence of barometric pressure" ... *Ibid.*, 219.

72 Like Redfield, William Ferrel ... Kutzbach, *The Thermal Theory of Cyclones*, 36.

73 Ferrel was also the first ... *Ibid.*, 39.

74 As late as 1822 meteorologist Robert Hare ... Robert Hare, "On the Gales Experienced in the Atlantic States of North America," *American Journal of Science* (1822), 355.

74 On a rainy Wednesday ... C. C. Gillespie, *The Montgolfier Brothers and the Invention of Aviation, 1738–1784* (Princeton: Princeton University Press, 1983), 2–4.

75 The first American airship ... James Alexander, *The Conquest of the Air* (London: S.W. Partridge & Co., 1902), 17.

76 After landing, Charles excitedly ... *Ibid.*, 22.

76 Said one early balloonist ... James Glaisher, ed., *Travels in the Air* (London: Richard Bentley & Son, 1871), 297.

76 "I dimly saw Mr. Coxwell" ... *Ibid.*, 53.

78 "[E]ndless variety and grandeur" ... *Ibid.*, 49.

78 As one aerologist put ... Alexander, *The Conquest of the Air*, 119.

78 Searching for this aerological ... A. Lawrence Rotch, *Sounding the Ocean of Air* (Ney York: E. & J. B. Young & Co., 1900), 100.

78 Lord Kelvin, one of the founders ... Kutzbach, *The Thermal Theory of Cyclones*, 46–48.

80 But in 1866, the Austrian meteorologist ... *Ibid.*, 58–61.

82 When theory and synoptic experience ... Chester Newton and Harriet Rodebush Newton, "The Bergen School Concepts Come to America," in M. A. Shapiro and S. Gronas, eds., *The Life Cycles of Extratropical Cyclones* (Boston: American Meteorological Society, in press).

CHAPTER FIVE: THE AVIATION AGE

83 "He had been trying to avoid" ... *New York Times* (4 September 1925), 3.

84 After the crash, blame ... *New York Times* (11 September 1925), 8, and (13 September 1925), 16.

85 Even the infamous *Hindenburg* ... J. Gordon Vaeth, "What Happened to the *Hindenburg*," *Weatherwise* (December 1990/January 1991), 315–323.

85 "I shook all the way" ... Carl Solberg, *Conquest of the Skies* (Boston: Little, Brown, 1979), 19.

86 One passenger remembered ... *Ibid.*, 127.

87 One man saw to it ... Patrick Hughes, "Francis W. Reichelderfer: Aerologists and Airdevils," *Weatherwise* (April 1981), 52.

87   "That put the fear of God into me" … *Ibid.*, 55.

88   One meteorologist later sneered … Tor Bergeron, "Synoptic Meteorology: An Historical Review," in Gösta H. Liljequist, ed., *Weather and Weather Maps* (Basel: Birkäuser Verlag, 1981), 453.

88   "Analysis of the weather map" … Francis W. Reichelderfer, "The Early Years," *Bulletin of the American Meteorological Society* (March 1970), 207.

89   "[T]here are hundreds of rules" … Jerome Namias, "The History of Polar Front and Air Mass Concepts in the United States—An Eyewitness Account," *Bulletin of the American Meteorological Society* (July 1983), 748.

89   Not surprisingly, Harrison was … Gordon D. Cartwright, "The Development of Aeronautical Meteorology: The Early Years," preprint for paper presented at the American Meteorological Society Annual Meetings, Dallas, Texas, 1995.

90   They represented a dying breed … Robert Marc Friedman, *Appropriating the Weather: Vilhelm Bjerknes and the Construction of a Modern Meteorology* (Ithaca, N.Y.: Cornell University Press, 1989), 21–28.

90   Better yet though, Scandinavia was … *Ibid.*, 33–34, 42–44.

92   The new theorem excited Ekholm … Kutzbach, *The Thermal Theory of Cyclones*, 159–160.

92   Ekholm's one-day crash course … Friedman, *Appropriating the Weather*, 37.

92   Bjerknes realized that his friends … *Ibid.*, 38.

92   Abbe soon sent Bjerknes … *Ibid.*, 38–39.

94   The circulation theorem was a hit … *Ibid.*, 34, 40–44.

94   "A physicist who goes into meteorology" … *Ibid.*, 48.

94   One day in 1902 … C.J.P. Cave, "The Great Days of Kite Flying," *Weather* (May 1947), 136

95   He told Nansen that … Friedman, *Appropriating the Weather*, 55.

95   For instance, upper-air charts … *Ibid.*, 66.

95   A Zeppelin traveling … *Ibid.*, 74.

95   "For the first time, we have made" … *Ibid.*, 102.

96   "British soldiers … do not go into action" … , *Ibid.*, 104.

97   "One often sat up the night" … Ralph Jewell, "Tor Bergeron's First Year in the Bergen School: Towards an Historical Appreciation," in *Weather and Weather Maps*, 473–474.

98   Occasionally, he would look in … *Ibid.*, 484.

98   "Jack really enjoys his work" … *Ibid.*, 481.

98   Furthermore, earlier convergence theories … Friedman, *Appropriating the Weather*, 130–132.

99   But none of the storm observations … *Ibid.*, 158–163.

100 "The warm air is victorious" … *Ibid.*, 188.

100 At first, the Bergen analysis maps … G.H. Liljequist, "Tor Bergeron," in *Weather and Weather Maps*, 413.

101 Furthermore, Bigelow correctly ascertained … Kutzbach, *The Thermal Theory of Cyclones*, 179.

104 Vilhelm Bjerknes wrote admiringly … Jewell, "Tor Bergeron's First Year in the Bergen School," 487.

107 The incident shook Bergeron … Liljequist, "Tor Bergeron," 419.

107   Practically no one had bothered … Namias, "The History of Polar Front and Air Mass Concepts in the United States," 734.

107   Some say he was compensating … Tor Bergeron, "The Young Carl-Gustaf Rossby," in Bert Bolin, ed., *The Atmosphere and the Sea in Motion* (New York: The Rockefeller Institute Press and Oxford University Press, 1959), 51.

107–108   "This boy … had an amazing" … *Ibid.*, 53.

108   Observations were also an issue … Newton and Newton, "The Bergen School Concepts Come to America," 7.

109   Local officials treated the enthusiastic … Horace R. Byers, "Carl-Gustaf Rossby, the Organizer," in *The Atmosphere and the Sea in Motion*, 56.

109   "Those who studied under [Rossby]" … *Ibid.*, 57.

110   "Edith had … a wonderful feel for symmetry" … H. Taba, "Interview: Jerome Namias," *WMO Bulletin* (July 1988), 160.

111   "This was the key" … H. Taba, "Interview: George P. Cressman," *WMO Bulletin* (October 1996), 317.

112   When he came back, his report … Gordon D. Cartwright and Charles H. Sprinkle, "A History of Aeronautical Meteorology: Personal Perspectives, 1903–1905," in Fleming, ed., *Historical Essays on Meteorology, 1919–1995*, 454.

112   One sounding balloon launched in England … D. H. Johnson, "The Jet Stream, Part 1," *Weather* (September 1953), 270.

113   Rossby's own department at Chicago … William A. Koelsch, "From Geo- to Physical Science: Meteorology and the American University, 1919–1945," in Fleming, ed., *Historical Essays on Meteorology, 1919–1995*, 530.

CHAPTER SIX: BREWING THE STORM

115   A storm with a sharp cold … Paul J. Kocin, Daniel H. Graf, and William E. Gartner, "Snow," *Weatherwise* (February/March 1995), 26, 28.

116   Patrick Michaels, a University of Virginia … William Claiborne, "Weathering the Sleeze Factor," *Washington Post* (11 February, 1994), A27.

116   The toppled trees … Roger Petterson (Associated Press), "Winter-Ravaged Regions Dig Out, Power Up as New Storm Threatens," *Louisville Courier-Journal* (13 February 1994), A2

118   Many early thinkers believed that clouds … Day and Ludlam, "Luke Howard and His Clouds," 460.

120   Aitken developed a special chamber … B. J. Mason, *Clouds, Rain and Rainmaking* (Cambridge: Cambridge University Press, 1962), 16–18.

120   Another Scot, C.T.R. Wilson … Marjory Roy, "Memorial Plaque to C. T. R. Wilson," *Weather* (July 1997), 224–225.

120   He was a skilled glassblower … E. C. Halliday, "Some Memories of Prof. C. T. R. Wilson, English Pioneer in Work on Thunderstorms and Lightning," *Bulletin of the American Meteorological Society* (December 1970), 1133–1135

120   In the Sahara Desert … Frank H. Ludlam, *Clouds and Storms: The Behavior and Effect of Water in the Atmosphere* (University Park and London: The Pennsylvania State University Press, 1980), 64–68.

121   The waves broke ... Duncan Blanchard, *From Raindrops to Volcanoes: Adventures with Sea Surface Meteorology* (Garden City, N.Y.: Doubleday, 1967), 57–59.

122   If the droplets can grow ... Mason, Clouds, Rain and Rainmaking, 74.

123   It took nearly half a century ... Patrick Hughes, "The Meteorologist Who Started A Revolution," *Weatherwise* (April/May 1994), 29.

124   At the age of 17 ... Liljquist, "Tor Bergeron," 409.

124   In 1922, between stints ... *Ibid.*, 415–417.

125   In fact, similar moist air ... Walter A. Lyons, Ph.D., *The Handy Weather Answer Book* (Detroit: Visible Ink Press, 1997), 204

126   Nearby, at Montague, New York ... Lee Grenci, "Montague versus the Capulets," *Weatherwise* (April/May 1997), 50.

126   In one legendary lake-effect ... Henry J. Cox, "South Chicago's Thanksgiving Day Snowstorm," *Weatherwise* (December 1972), 286–287

127   The worst East Coast snow storms ... Paul J. Kocin and Louis W. Uccellini, *Snowstorms along the Northeastern Coast of the United States, 1955–1985* (Boston: American Meteorological Society, 1990), 33–35.

128   In January 1998, however ... Lee Grenci, "Glazed Over," *Weatherwise* (May/June 1998), 50.

128   A New York state trooper ... Jim Yardley, "Northern New York Battles Ice Damage," *New York Times* (12 January 1998), B6.

129   Early on scientists homed ... B. J. Mason, "Recent Developments in the Physics of Rain and Rainmaking," *Weather* (March 1959), 81.

130   Many rainy sheets of cloud ... Robert A. Houze, Jr., *Cloud Dynamics* (San Diego: Academic Press, 1993), 205–217.

131   The theory was far better ... Roscoe R. Braham, Jr., "Formation of Rain: A Historical Perspective," in Fleming, ed., *Historical Essays on Meteorology, 1919–1995*, 195–197.

131   In the 1940s ... *Ibid.*, 199.

131   A rain cloud can last an hour ... Frank H. Ludlam, "Artificial and Natural Shower Formation," *Weather* (July 1952), 199–204.

131   A young researcher in England, *Ibid.*, 202–204.

133   Langmuir and his assistant ... Vincent J. Schaefer, "Halley's Comet, An Arrowhead, and ... Lots of Luck," *Weatherwise* (December 1985), 304–308.

133   The meteorologists who spend ... David Thurlow, "To Join the Club, 'Get Small,'" *Weatherwise* (April/May 1995), 54.

135   Langmuir got a bit carried away ... "Artificial Weather Made," *Science News Letter* (1 November 1947), 277, and "Rain-Maker Proposes Eliminating Lightning," *Science News Letter* (20 November 1948), 328.

136   In 1949, Weather Bureau Chief ... "Rain-Making, Pro and Con," *Science News Letter* (11 February 1950), 87.

136   The project used an old B-17 ... Joe Chew, *Storms above the Desert: Atmospheric Research in New Mexico, 1935–1985* (Albuquerque: University of New Mexico Press, 1987), 36.

136   Heavy snows that year ... "Rain-Making Is Assailed," *New York Times* (13 June 1951), 32

136   At the next meeting ... "Seeding May Reduce Rain," *Science News Letter* (9 February 1952), 83–84.
138   In 1966 the Air Force ... Bates and Fuller, *America's Weather Warriors*, 230–231.
138   Department of Interior rainmakers ... Anthony Ripley, "Funerals Start in Dakota Flood," *New York Times* (14 June 1972), II:1.

CHAPTER SEVEN: TERRIBLE UPS AND DOWNS

141   When Lieutenant Colonel William H. Rankin ... William H. Rankin, "I Rode the Thunder," *Saturday Evening Post* (15 October 1960), 24–25.
142   The idea was nearly forgotten until ... , A. E. Slater, 'The 'Mystery' of Soaring Flight," *Weather* (September 1955), 298–303
143   In fact, tropical cumulus clouds are ... Joanne Starr Malkus, "Aeroplane Studies of Trade-Wind Meteorology," *Weather* (October 1953), 297
144   Stommel compared rising cloudy air ... Houze, *Cloud Dynamics*, 226–227.
144   By 1950 Horace Byers ... Horace Byers, "Structure and Dynamics of the Thunderstorm Part II," *Weather* (August 1949), 247
144   Scorer demonstrated how ... Frank H. Ludlam, "Large Air Bubbles in Water," *Weather* (June 1954), 169
147   Byers later compared ... Byers, "Structure and Dynamics of the Thunderstorm, Part I," *Weather* (July 1949), 221.
149   Even so, the cloud must make ... Lyons, *Handy Weather Answer Book*, 151.
149   At peak, each lightning ... *Ibid.*, 146–147.
150   "We took it for granted" ... Sir George C. Simpson, "Wilson's Theory of the Normal Electrical Field, Part II," *Weather* (May 1949), 134–140.
150   His observations with a stopwatch ... Chew, *Storms Above the Desert*, 9.
151   Some scientists are hoping to develop ... Hazel Muir, "Striking Back at Lightning," *New Scientist* (7 October 1995), 26–30.
151   The newer picture of storms ... Earle R. Williams, "The Electrification of Thunderstorms," *Scientific American* (November 1988), 91–94.
153   This once was a serious problem ... Martin A. Uman, *All about Lightning* (New York: Dover Publications, 1986), 30.
155   In Chicago Heights ... Michele L. Norris and Liz Sly, "Lightning fells 30 in Chicago Heights," *Chicago Tribune* (30 July 1997), 1.
155   Chicagoans got another taste ... Hugh Dollios, "Lightning Injures 11 as Bolt Hits at Plant," *Chicago Tribune* (26 January 1990), 10.
155   And just to prove ... Lyons, *Handy Weather Answer Book*, 145.
156   Most hail larger than a marble ... *Ibid.*, 133.
156   Generally, each layer might form ... Narayan R. Gokhale, *Hailstorms and Hailstone Growth* (Albany: State University of New York Press, 1975), 137–153.
157   By the time the hail ... Patrick Hughes and Richard Wood, "Hail: The White Plague," *Weatherwise* (April/May 1993), 18.
157   Five glider pilots bailed ... *Ibid.*, 19.
158   In one wind tunnel test ... Gokhale, *Hailstorms and Hailstone Growth*, 162–168.

158 Such observations prompted Soviet ... William R. Cotton, *Storms* (Fort Collins, Colo.: *ASTeR Press, 1990), 72–73.

158 The technique was used ... H.W. Sansom, "The Use of Explosive Rockets to Suppress Hail in Kenya," *Weather* (March 1966), 86–91.

160 "One may say that heretofore" ... Byers, "The Structure and Dynamics of the Thunderstorm Part II," 248.

160 "We spent millions" ... Stanley David Gedzelman, "Mysteries in the Clouds," *Weatherwise* (June/July 1995), 57.

161 One night in 1947 ... Tetsuya Theodore Fujita, *Memoirs of an Effort to Unlock the Mystery of Severe Storms During the 50 Years, 1942–1992* (Chicago: Department of Geophysical Sciences, University of Chicago, 1992), 187–190

161 Fujita's colleagues thanked him ... *Ibid.*, 197–198.

162 At first lightning caught ... Lawrence Van Geldner, "109 Feared Dead As Jet Falls Near Kennedy During A Storm," *New York Times* (25 June 1975), 1, 20–21.

164 Fujita's microburst studies ... Fujita, *Memoirs*, 103–128.

165 One microburst struck near ... Lyons, *Handy Weather Answer Book*, 116.

165 A Delta wide-body jet ... William G. Laynor, "Summary of Wind Shear Accidents and Views About Prevention," *Wind Shear* (Warrendale, Penn.: Society of Automotive Engineers, 1986), 7–8.

165 On Highway 114 ... Doug J. Swanson, "Witnesses Describe Horror," *Dallas Morning News* (3 August 1985), 1A, 38A.

166 Dallas-Fort Worth actually had ... Dan Malone, "Jet May Have Received No Storm Alert," *Dallas Morning News* (4 August, 1985), 1A, 15A.

168 The low-level outflow ahead ... Robert A. Houze Jr. et al., "Interpretation of Midlatitude Mesoscale Convective Systems," *Bulletin of the American Meteorological Society* (June 1989), 608–611.

169 These storms often originate ... Cotton, *Storms*, 97–112.

170 Airlines must watch ... Robert A. Maddox and J. Michael Fritsch, "A New Understanding of Thunderstorms: The Mesoscale Convective Complex," *Weatherwise* (June 1984), 133.

170 A community caught under an MCC ... J. M. Fritsch et al., "The Contribution of Mesoscale Convective Weather Systems to the Warm-Season Precipitation in the United States," *Journal of Climate and Applied Meteorology* (October 1986), 1333–1345.

171 "The first velocity images" ... John W. Cannon, "A Ringside Seat," *Weatherwise* (August/September 1996), 14.

172 Campers in the mountains ... Mace Bentley, "A Midsummer's Nightmare," *Weatherwise* (August/September 1996), 11–13.

172 Why some MCSs develop derechos ... Cotton, *Storms*, 93–94.

172 A major derecho crossed Nebraska ... Mace L. Bentley and Stonie R. Cooper, "The 8 and 9 July 1993 Nebraska Derecho: An Observational Study and Comparison to the Climatology of Related Mesoscale Systems," *Weather and Forecasting* (September 1997), 678–688.

174 A backdoor cold front ... Lee R. Hoxit et al., "Disaster by Flood," in Edwin Kessler, ed., *The Thunderstorm in Human Affairs* (Norman and London: University of Oklahoma Press, 1983), 20–34.

175 Now more than twice ... John F. Henz et al., "The Big Thompson Flood of 1976 in Colorado," *Weatherwise* (December 1976), 285.

175 "The whole mountainside is gone" ... Hoxit et al., "Disaster by Flood," 27.

175 More than 1,000 people ... Bill Myers, "Rain, Fog Delay Canyon Flood Rescue," *The Denver Post* (2 August 1976), 3.

176 A solitary storm forged ahead ... Jeffrey Weiss, "Supercell Nails Bull's-Eye: Dallas and Fort Worth," *Dallas Morning News* (7 May 1995), A28.

176 A severe storm like it ... Michael Saul and J. Lynn Lunsford, "Damage from Hail Grounds Nearly 100 D/FW Planes," *Dallas Morning News* (1 May 1995), A1.

177 "Like a rockslide" ... J. Lynn Lunsford, "Visitors at FW Mayfest Forced to Flee, Seek Shelter," *Dallas Morning News* (6 May 1995), A1.

177 A teenager rescued ... Nora Lopez, "In Air, on Land, in Water, Storm Disrupts Lives," *Dallas Morning News* (7 May 1995), 30A.

177 When floodwaters covered ... Randy Lee Loftis and Nora Lopez, "16 Deaths Blamed on Storm," *Dallas Morning News* (7 May 1995), 1A, 29A.

CHAPTER EIGHT: MONSTERS OF THE PRAIRIE

179 The young have mostly left ... Jon Jeter, "Many Unsure About Tornado Town That Was Already Dying," *Washington Post* (2 June 1998), A2.

179 Face-to-face with the planet's ... Pam Belluck, "In Ruins of Tornado, Holding on to Simple Things," *New York Times* (2 June 1998), A14.

181 Flora extensively quoted ... H.T. Harrison, *Certain Tornado and Squall Line Features* (United Airlines Meteorology Circular No. 36), quoted in Snowden D. Flora, *Tornadoes of the United States* (Norman: University of Oklahoma Press, 1953), 12.

182 Or how a Scottsbluff ... Flora, *Tornadoes of the United States*, 17.

182 The worst in American history ... Peter S. Felknor, *The Tri-State Tornado: The Story of America's Greatest Tornado Disaster*, (Ames: Iowa State University Press, 1992).

183 Will Keller, a farmer ... Flora, *Tornadoes of the United States*, 10–11.

183 Hall and his family ... Roy S. Hall, "Inside a Texas Tornado," *Weatherwise* (June 1951, reprinted in April 1987), 72–75.

184 In his study of tornadoes ... Flora, *Tornadoes of the United States*, 6–7.

184 Even television seemed ... "The Weller Method of Tornado Detection," *Weatherwise* (June 1970), 121–123.

185 Strictly speaking, meteorologists ... William R. Corliss, *Handbook of Unusual Natural Phenomena* (New York: Gramercy Books, 1995), 212–215.

185 Stirling Colgate, an astrophysicist ... Howard B. Bluestein and Joseph H. Golden, "A Review of Tornado Observations," in Christopher R. Church et al., ed., *The Tornado, Its Structure, Dynamics, Prediction, and Hazards* (Washington, D.C.: American Geophysical Union, 1993), 332–333.

187 The bodies of turkeys ... B. Vonnegut, "Chicken Plucking as Measure of Tornado Wind Speed," *Weatherwise* (October 1975), 217.

189 Fujita also studied a famous ... Fujita, *Memoirs*, 27–28.

189 The strongest tornado … Polk Laffoon IV, *Tornado* (New York: Harper & Row, 1975).

189 Near Branchville, Indiana, a school bus … "Twisters Whipped Indiana's Small Towns," *Louisville Courier-Journal* (4 April 1974), 6.

189 A man driving home … Billy Reed, "To Hard-Hit Residents of Northfield the Tornado Seemed to Last Forever," *Louisville Courier-Journal* (4 April 1974), A4.

191 This scale reveals what engineers … Joseph E. Minor, "Effects of Wind on Buildings," in Kessler, ed., *The Thunderstorm in Human Affairs*, 89–104.

192 For instance the Jarrell … Long T. Phan and Emil Simiu, *The Fujita Tornado Intensity Scale: A Critique Based on Observations of the Jarrell Tornado of May 27, 1997*, NIST Technical Note 1426, (Washington, D.C.: Government Printing Office, 1998).

193 *Birmingham News* reporter … Russell Hubbard, "Path of Horror Ends Just 3 Miles from City's Heart," *Alabama Live/Birmingham News* Online (10 April 1998).

193 One man spoke for … John Archibald and Phil Pierce, "Area's Worst Tornado Forever Mars Land, Lives," *Alabama Live/Birmingham News* Online (10 April 1998).

194 The tornado was often transparent … Robert F. Abbey Jr. and T. Theodore Fujita, "The Tornado Outbreak of 3–4 April 1974," in Kessler, ed., *The Thunderstorm in Human Affairs*, 52–55.

195 Johannes Letzmann, a pioneering … Richard E. Peterson, "Johannes Letzmann: A Pioneer in the Study of Tornadoes," *Weather and Forecasting* (March 1992), 166–184.

196 Ward started out as … Gene Rhoden, "Storm Pioneer: A Biography of Neil B. Ward," *Storm Track* (30 November 1990).

196 Using tornado models, meteorologists … Christopher R. Church and John T. Snow, "Laboratory Models of Tornadoes," in Church ed., *The Tornado*, 282–284.

197 Benjamin Franklin probably launched … Benjamin Franklin (letter of 25 August 1775) quoted in John Pearl, "Founding Chaser?" *Weatherwise* (August 1995), 6.

198 Hoadley caught storm fever … Tim Marshall, "Chase Fever: The Early Years," *Storm Track* (31 January 1987).

198 In one editorial, Hoadley … David Hoadley, "Why Chase Tornadoes?" *Storm Track* (31 March 1982).

199 Marshall points out that … Tim Marshall, "A Passion for Prediction," *Weatherwise* (April/May 1995), 24–25.

199 "He is doing too much brain work" … Joseph G. Galway, "J.P. Finley: The First Severe Storms Forecaster (Part 1)," *Bulletin of the American Meteorological Society* (November 1985), 1390.

200 Hinrichs believed Finley … Joseph G. Galway, "J.P. Finley: The First Severe Storms Forecaster (Part 2)," *Bulletin of the American Meteorological Society* (December 1985), 1506–1510.

202 Meanwhile, the Signal Service … Joseph G. Galway, "Early Severe Thunderstorm Forecasting and Research by the United States Weather Bureau," *Weather and Forecasting* (December 1992), 564–585

204 "We could only pray" ... Robert C. Miller, *Events Leading to the First Operational Tornado Forecast*, (unpublished manuscript of 1948 transcribed by Charlie A. Crisp, distributed on-line by NOAA/NSSL, 1998).

205 Weather Bureau storm researcher ... Fujita, *Memoirs*, 7–10.

206 Then, in 1948, Edward Brooks ... Edward M. Brooks, "The Tornado Cyclone," *Weatherwise* (April 1949), 32–33.

206 But because of all the pictures ... Bluestein and Golden, "A Review of Tornado Observations," 319.

206 Horace Byers heard about ... Fujita, *Memoirs*, 21–26.

208 Not long after Fargo ... F.H. Ludlam, *Clouds and Storms: The Behavior and Effect of Water in the Atmosphere*, 216–220.

208 While Browning was looking ... Severe Storms Research Group of St. Louis University, "F.C. Bates' Thoughts on Severe Thunderstorms," *Bulletin of the American Meteorological Society* (June 1970), 481–489.

211 Joseph Klemp of the National ... Richard Rotunno, "Supercell Thunderstorm Modeling and Theory," in C.R. Church ed., *The Tornado*, 57–73.

213 Hunting for severe weather ... Alan Moller et al., "Field Observations of the Union City Tornado in Oklahoma," *Weatherwise* (April 1974), 68–77.

214 As one scientist put it ... Ralph Donaldson Jr., "Observations of the Union City Tornadic Storm by Plan Shear Indicator," *Monthly Weather Review* (January 1978), 45.

215 With chasers on the ground ... Leslie R. Lemon et al., "Tornadic Storm Airflow and Morphology Derived from Single Doppler Radar Measurements," *Monthly Weather Review* (January 1978), 48–61.

215 "I'd heard rumors that some people" ... Howard B. Bluestein (interview with Jeffrey Rosenfeld), "Spin Doctor," *Weatherwise* (April/May 1996), 19–20.

216 On April 26, 1991, Bluestein ... Howard B. Bluestein, "Riders on the Storm," *The Sciences* (March/April 1995), 26–28.

216 "I turned around to talk" ... *Ibid.*, 28.

217 In 1950 an airline pilot ... J.P. Henderson, "Waterspouts on Lake Victoria," *Weather* (September 1950), 326–328.

217 One of the chasers ... Joseph H. Golden, "Waterspouts at Lower Matecumbe Key, Florida, 2 September 1967," *Weather* (March 1968), 103–114.

217 The cores had updrafts ... Bluestein and Golden, "A Review of Tornado Observations," 322–323.

218 A computer model by Bruce Lee ... Daniel Pendick, "Virtual Vortex: Landspout in a Box," *Weatherwise* (May/June 1998), 24–31.

220 The Newcastle storm was ... Roger M. Wakimoto and Nolan T. Atkins, "Observations of the Origins of Rotation in the Newcastle Tornado During VORTEX '94," *Monthly Weather Review* (March 1996), 384–407.

220 ELDORA made many fascinating ... Roger M. Wakimoto et al., "ELDORA Observations during VORTEX '95," *Bulletin of the American Meteorological Society* (July 1996), 1465–1481.

221 In VORTEX, DOW resolved ... Joshua Wurman et al., "Fine-Scale Doppler Radar Observations of Tornadoes," *Science* (21 June 1996), 1774–1777.

221 Wurman's team was on hand ... Joshua Wurman, interview with author, 12 November 1996.

CHAPTER NINE: THE ULTIMATE STORM

224 Slow to respond to the threat ... Larry Rohter, "Hurricane Leaves Widespread Damage and at Least 110 Dead," *New York Times* (24 September 1998), A14.

224 Some people had no choice ... Joseph B. Treaster and Raymond Hernandez, "Making Do with Supplies, Dominican Republic Struggles to Rebuild," *New York Times* (28 September 1998), A12.

225 The attitude seemed naive ... Garry Pierre-Pierre, "After the False Alarms, Haitians Fail to Prepare," *New York Times* (28 September 1998), A14.

225 Nor does any other storm ... Robert H. Simpson and Herbert Riehl, *The Hurricane and Its Impact* (Baton Rouge and London: Louisiana State University Press, 1981), 35.

226 Such work could be performed ... Robert H. Simpson, "Hurricanes," *Scientific American* (June 1954), 32.

227 In 1892, Ralph Abercromby ... J.M. Walker, "Pre-1850 Notions of Whirlwinds and Tropical Cyclones," *Weather* (December 1989), 483–484.

228 A flock of sea gulls ... Jay Barnes, "Creatures in the Storm," *Weatherwise* (September/October 1998), 27–28.

228 One mariner who endured ... Tannehill, *The Hurricane Hunters*, 78–79.

228 The Confederate ship *Alabama* ... Marjory Stoneman Douglas, *Hurricane* (New York: Rinehart, 1958).

228 Edward R. Morrow, the CBS newsman ... Tannehill, *The Hurricane Hunters*, 200.

230 Otherwise the air might cool ... Roger A. Pielke, *The Hurricane* (London and New York: Routledge, 1990), 41.

230 He heard that a hurricane ... H. Taba, "The *Bulletin* Interviews: Professor E.H. Palmén," *WMO Bulletin* (April 1981), 98.

230 In fact, the leading hurricane expert ... Dick DeAngelis, "The Hurricane Priest," *Weatherwise* (October 1989), 256–257.

230 At the appearance of the actual ... Tannehill, *Hurricanes* (Princeton, N.J.: Princeton University Press, 1957), 97.

232 Isaac Cline, the Weather Bureau chief ... Patrick Hughes, "The Great Galveston Hurricane," *Weatherwise* (August 1990), 190–198.

234 One of the highest storm surges ... A.K. Sensarma, "The Great Bengal Cyclone of 1737—An Enquiry into the Legend," *Weather* (March 1994), 90–96.

235 And on April 29, 1991, Bangladeshis ... Jawador Rahman, *Living with Cyclone: Strategies for People Against the Fury of Nature* (Bangladesh: Center for Sustainable Development and Christian Commission for Development in Bangladesh, 1993).

236 One survivor swore ... Gary Dean Best, *FDR and the Bonus Marchers* (New York: Crown Publishers Inc., 1967).

237 The life-saving satellites ... Mark DeMaria, "A History of Hurricane Forecasting for the Atlantic Basin, 1920–1995," in Fleming ed., *Historical Essays on Meteorology, 1919–1995*, 283.

237 The dream of weather surveillance ... James F.W. Purdom and W. Paul Menzel, "Evolution of Satellite Observations in the United States and Their

Use in Meteorology," in Fleming ed., *Historical Essays on Meteorology, 1919–1995*, 103–117.

239   The plane was "tossed about" … Tannehill, *The Hurricane Hunters*, 99.

239   Flying into a 1945 hurricane … *Ibid.*, 137.

239   The water leaked … *Ibid.*, 145.

240   "Shafts of supercooled water" … *Ibid.*, 257.

240   "The plane was simply too loaded" … *Ibid.*, 259.

240   With radar guidance … Robert H. Simpson, "Studying Hurricanes from the Inside," *Weatherwise* (August 1948, reprinted in January/February 1998), 47.

241   The NOAA forecast of Georges' … Larry Rohter, "Killer Hurricane Takes Property, But Apparently No Lives in Keys," *New York Times* (26 September 1998), A1.

242   In the 1930s, Gordon Dunn … Robert W. Burpee, "Gordon E. Dunn: Preeminent Forecaster of Midlatitude Storms and Tropical Cyclones," *Weather and Forecasting* (December 1989), 575.

242   Despite nearly unvarying conditions … Joanne S. Malkus, "Tropical Weather Disturbances—Why Do So Few Become Hurricanes?" *Weather* (March 1958), 75–89.

243   It was a banner year … William M. Gray et al., *Summary of Atlantic Tropical Cyclone Activity and Verification of Authors' Seasonal Prediction* (Fort Collins: Colorado State University, 1995).

244   The wave probably throws … Malkus, "Tropical Weather Disturbances," 86.

244   "This astonishing temperature gradient" … Simpson, "Hurricanes," 34.

246   Calculations showed the storm … Robert C. Sheets and Noel E. La Seur, "Project Stormfury: Present Status—Future Plans," *WMO-Bulletin* (January 1979), 17–23.

248   "It was exactly the case" … Hugh Willoughby, interview with author, 16 January 1998.

248   In the hurricane, according to recent … Michael T. Montgomery, "Vortex Intensification by Convectively Forced Vortex Rossby Waves," *Preprints, Symposium on Tropical Cyclone Intensity Change* (Boston: American Meteorological Society, 1998), 21.

249   As helpful as the scale is … Lixion A. Avila, "Forecasting Tropical Cyclone Intensity Changes: An Operational Challenge," *Preprints, Symposium on Tropical Cyclone Intensity Change* (Boston: American Meteorological Society, 1998), 1.

250   In 1997, Hurricane Linda … Robert Henson, "The Intensity Problem," *Weatherwise* (September/October 1998), 23.

250   But with only five feet … Stephen K. Boss and A. Conrad Neumann, "Hurricane Andrew on the Northern Great Bahama Bank," *Journal of Coastal Research* (Spring 1995).

250   At the same time, Opal met … Lance F. Bosart et al., "Environmental Influences on the Rapid Intensification Stage of Hurricane Opal (1995) over the Gulf of Mexico," *Preprints, Symposium on Tropical Cyclone Intensity Change* (Boston: American Meteorological Society, 1998), 105–112.

251   The building itself began to shake … Jack Williams, "Tracking Andrew from Ground Zero," *Weatherwise* (December 1992/January 1993), 8–12.

251 A photographer who rode out ... Warren Faidley, "In Harm's Way," *Weatherwise* (December 1992/January 1993), 16.

251 Oddly the disaster seemed magnified ... Roger M. Wakimoto and Peter G. Black, "Damage Survey of Hurricane Andrew and Its Relationship to the Eyewall," *Bulletin of the American Meteorological Society* (February 1994), 189–200.

252 Seven convective cells spun up ... Hugh E. Willoughby and Peter G. Black, "Hurricane Andrew in Florida: Dynamics of a Disaster," *Bulletin of the American Meteorological Society* (March 1996), 543–549.

252 Researchers encountered a similar ... Robert D. Fletcher et al., "Typhoon Ida from Above," *Weatherwise* (June 1961), 102–105.

254 Major advances in such modeling ... Joanne and Robert Simpson, interview with author, 14 January 1998.

CHAPTER TEN: AN AWESOME CHAOS

259 The ever-provocative lightning ... Otha H. Vaughan Jr. and Bernard Vonnegut, "Lightning to the Ionosphere?" *Weatherwise* (April 1982), 70–72.

259 Attitudes changed with one ... Walter Lyons, "Sprites, Elves and Blue Jets," *Weatherwise* (August/September 1997), 20.

261 They were amazed to discover ... *Ibid.*, 21.

264 These winter Arctic Ocean storms ... Steven Businger, "Arctic Hurricanes," *American Scientist* (January-February 1991), 18–33.

265 Satellite images also have revealed ... J.A. Ernst and M. Matson, "A Mediterranean Tropical Storm?" *Weather* (November 1982), 332–337.

265 Off the coasts of the United States ... Robert Henson, "Hurricanes in Disguise," *Weatherwise* (December 1995/January 1996), 12–17.

265 NOAA scientist Mel Shapiro ... Paul J. Neiman and M.A. Shapiro, "The Life Cycle of an Extratropical Marine Cyclone—Part I: Frontal-Cyclone Evolution and Thermodynamic Air-Sea Interaction," *Monthly Weather Review* (August 1993), 2153–2176.

266 A University of Wisconsin meteorologist ... Jonathan E. Martin, "The Structure and Evolution of a Continental Winter Cyclone—Part I: Frontal Structure and the Occlusion Process," *Monthly Weather Review* (February 1998), 314.

267 "Unfortunately," he admitted ... Clifford F. Mass, "Synoptic Frontal Analysis: Time for a Reassessment?" *Bulletin of the American Meteorological Society* (March 1991), 348.

269 A more likely theory ... Frederik Nebeker, *Calculating the Weather: Meteorology in the 20th Century* (San Diego: Academic Press, 1995), 137.

271 The computer often sputtered ... Stanley David Gedzelman, "Chaos Rules," *Weatherwise* (August/September 1994), 21–28.

273 "If an adequate definition" ... Tannehill, *The Hurricane Hunters*, 200.

Index

Numbers in italics indicate photos. U.S. storms and other local weather characteristics are indexed by state and/or region.